Life And Her Children

LIFE AND HER CHILDREN

THE NEW YORK
PUBLIC LIBRARY

ASTOR LENOX
TILDEN FOUNDATIONS

LIFE IN THE DEEP SEA.
(for description see list of illustrations)

LIFE

AND HER CHILDREN

GLIMPSES OF ANIMAL LIFE

FROM THE AMŒBA TO THE INSECTS

BY

ARABELLA B. BUCKLEY

AUTHOR OF
"THE FAIRYLAND OF SCIENCE," "A SHORT HISTORY OF NATURAL SCIENCE," ETC.

WITH UPWARDS OF ONE HUNDRED ILLUSTRATIONS.

'He prayeth best who loveth best
All things both great and small;
For the dear God who loveth us,
He made and loveth all.'
—COLERIDGE.

NEW YORK:
D. APPLETON AND COMPANY,
1, 3, AND 5 BOND STREET.
1885.

> 'His parent hand,
> From the mute shell-fish gasping on the shore,
> To men, to angels, to celestial minds,
> For ever leads the generations on
> To higher scenes of being; while supplied
> From day to day with his enlivening breath,
> Inferior orders in succession rise
> To fill the void below.'
>
> AKENSIDE.—*Pleasures of the Imagination.*

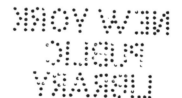

PREFACE.

THE plan of this work is so fully explained in the Introductory Chapter that but little preface is needed. Its main object is to acquaint young people with the structure and habits of the lower forms of life; and to do this in a more systematic way than is usual in ordinary works on Natural History, and more simply than in text-books on Zoology.

For this reason I have adopted the title "Life and her Children," to express the family bond uniting all *living* things, as we use the term "Nature and her Works," to embrace all organic and inorganic phenomena; and I have been more careful to sketch in bold outline the leading features of each division, than to dwell upon the minor differences by which it is separated into groups.

I have made use of British examples in illustration wherever it was possible, and small specimens of most

vi   PREFACE.

of the marine animals figured may be found upon our coasts at low tide.

In conclusion, I wish to express my great obligation to Mr. R. Garnett of the British Museum, for his most kind assistance in finding works of reference on the special subjects; and to many men of science, especially Mr. Lowne, F.R.C.S., and Mr. Haddon, Demonstrator of Comparative Anatomy at Cambridge, for their valuable criticisms on the proof-sheets.

The Illustrations of the marine animals have been drawn by Dr. Wild, artist of the 'Challenger' Expedition, and those of the insects by Mr. Edwin Wilson, to both of whom my thanks are due for the care and assiduity with which they have carried out my instructions.

ARABELLA B. BUCKLEY.

LONDON, *November* 1880.

CONTENTS.

CHAPTER I.
LIFE AND HER CHILDREN . . . . PAGE 1

CHAPTER II.
LIFE'S SIMPLEST CHILDREN, HOW THEY LIVE, AND MOVE, AND BUILD 14

CHAPTER III.
HOW SPONGES LIVE 33

CHAPTER IV.
THE LASSO-THROWERS OF THE PONDS AND OCEANS . . 50

CHAPTER V.
HOW STAR-FISH WALK AND SEA-URCHINS GROW . . 77

CHAPTER VI.
THE MANTLE-COVERED ANIMALS, AND HOW THEY LIVE WITH HEADS AND WITHOUT THEM 103

CHAPTER VII.
THE OUTCASTS OF ANIMAL LIFE, AND THE ELASTIC-RINGED ANIMALS BY SEA AND BY LAND 135

CHAPTER VIII.

THE MAILED WARRIORS OF THE SEA, WITH RINGED BODIES AND JOINTED FEET 153

CHAPTER IX.

THE SNARE-WEAVERS AND THEIR HUNTING RELATIONS . 178

CHAPTER X.

INSECT SUCKERS AND BITERS WHICH CHANGE THEIR COATS BUT NOT THEIR BODIES 201

CHAPTER XI.

INSECT SIPPERS AND GNAWERS WHICH REMODEL THEIR BODIES WITHIN THEIR COATS 233

CHAPTER XII.

INTELLIGENT INSECTS WITH HELPLESS CHILDREN, AS ILLUSTRATED BY THE ANTS 269

ILLUSTRATIONS.

PLATE I. LIFE IN THE DEEP SEA.—1. Sea-Lily, *Pentacrinus asteria*. 2. Sponge, *Euplectella aspergillum*. 3. Coral, *Lophohelia prolifera*. 4. Sea-Urchin, *Echinus elegans*. 5. Basket-fish, *Asterophyton linkii*. 6. Sea-cucumber, *Cladodactyla crocca*. 7. Jelly-fish, *Pelagia noctiluca*. 8. Pteropod, *Clio pyramidata*.

FRONTISPIECE.

PLATE II. INSECT LIFE.—1. Ant-Lion in its pit with whole insect shown above. 1a. Ant-lion flying. 2. Tiger-Beetle, *Cicindela campestris*. 3. Bombardier Beetle, *Brachinus crepitans*. 4. Burying Beetle, *Necrophorus interruptus*. 5. Cock-tail Beetle, *Staphylinus olers*. 6. Swallow-tailed Butterfly, *Papilio Machaon* . . . . *to face p.* 135

ILLUSTRATIONS IN THE TEXT.

	PAGE
The Thread-slime, *Protogenes*	16
The Finger Slime, *Protamœba*	18
Infusoria, *Monas, Noctiluca, Vorticella*	20
Foraminifera, *Miliola, Globigerina*, etc.	23
Jelly-body of a Miliolite	24
Nummulite, showing the chambers	27
Sun-Slime, *Physematium*	29
The Flint Shells of Radiolariæ or Polycistinæ . . .	30
A fragment of bath-sponge magnified	34
A British sponge found at Brighton	37
A Sponge-egg and the young sponge swimming . .	38
Development of a young English sponge . . .	39
Section of a bath-sponge, showing the chambers . .	41

ILLUSTRATIONS.

	PAGE
A Lime-sponge with the living flesh	44
Spicules of flint-sponges	45
Sarcode or flesh of a flint-sponge with the spicules	46
Venus' Basket. The skeleton of a flint-sponge	47
Cup-sponge growing at the bottom of the sea	48
Fresh-water hydra hanging from duckweed	51
Lasso-cells of the Hydra and Sea-Anemone	53
The Sea-Oak, *Sertularia*	56
The Campanulina, an animal-tree giving off jelly-bells	59
A jelly-fish, *Chrysoara hysocella*	63
The childhood of the same jelly-fish	65
Section of a Sea-Anemone showing its parts	67
Group of Sea-Anemones	68
Growth of Red Coral	72
A section of a piece of Red Coral	73
A piece of White Coral	74
The Devonshire Cup-Coral, *Caryophyllium Smythii*	75
The infancy of a Feather Star-fish	78
The infancy of a Brittle Star-fish	79
The infancy of the common Star-fish	80
The infancy of a Sea-Urchin	81
The infancy of a Sea-Cucumber	82
The common five-fingered Star-fish and the Brittle Star-fish	84
Section of the centre and of one ray of a Star-fish	85
The life of a Feather-Star	90
A Sea-Urchin walking on a rock	94
A Sea-Urchin with its spines rubbed off	95
An Oyster (Ostrea edulis) lying in the shell, showing the gills, mouth, etc.	108
A group of headless Mollusca, Cockle, Mussel, Scallop, and Razor-fish	111
Mollusca with heads, Vegetable-feeders, Limpet and Periwinkle	114
The anatomy of a Periwinkle	115
Flesh-feeding Molluscs, Whelk and Cowry	118
Garden Snail, Great Grey Slug, and Testacella	122
Naked-gilled Mollusca or sea-slugs, *Doris pilosa* and *Eolis coronata*	124
Oceanic Mollusca, *Ianthina, Carinaria*, and a Pteropod	125
Octopus shooting backwards through the water	127
The Mother Argonaut floating in the water	132
Land-Leeches of Ceylon racing to attack some creature	143
Section of a Leech showing the nervous system	144

ILLUSTRATIONS.

	PAGE
A group of fixed Sea-Worms—*Serpula, Terebella,* and *Spirorbis*	148
Active Sea-worms, *Aphrodite* or sea-mouse, and *Nereis*	151
A group of jointed-footed animals, *Arthropoda,* showing their ringed bodies	155
The Common Prawn	160
Sandhopper (*Talitrus*) and Skeleton Shrimp (*Caprella*)	163
Section of a prawn; and forepart with carapace removed showing the branchiæ	164
Metamorphoses of a Crab	167
Hermit-crabs in and out of the shell	170
Floating Barnacles, *Lepas,* with a bank of fixed Acorn-Barnacles, *Balanus*	174
Development of the Acorn Barnacle	176
A Scorpion with a Cricket in its claws	179
The parts of a Spider	183
Web of the garden spider	185
Nest of a trap-door spider	192
A hunting spider with a bag of eggs	195
The Water-Spider, *Argyroneta aquatica*	197
Aphides or plant-lice, with a grub feeding on them	202
The Cuckoo-spit insect (*Aphrophora spumaria*)	206
The Water-Measurer (*Gerris*) and Water-Boatman (*Notonecta*)	208
The large Green Grasshopper, its changes and its egg-laying	211
Spiracle and breathing-tube of an insect	212
Cockroaches—Young, male, and female, with egg-case	216
May-flies (*Ephemera*) and Caddis-flies (*Phryganea*) with their grubs	220
Dragon-fly, with the grub and the insect emerging	223
Section of an insect's eye	224
African Termites—king, queen, worker, and soldier	226
Queen Termite cell with queen within	228
Butterfly's head, caterpillar's head and cushion-foot, a butterfly's egg	237
Caterpillar, chrysalis, and perfect insect, of the Tortoise-shell butterfly	239
Caterpillar and chrysalis of Cabbage Butterfly	243
Caterpillar, cocoon, and moth of the six-spot Burnet	246
Psyche graminella, caterpillar and moth	247
Clothes-Moth with grub and pupa	249
Cockchafer grub, cocoon, and beetle	252
The Nut-weevil, maggot and beetle	254

		PAGE
Carnivorous Beetles, *Dyticus marginalis*, and the whirligig beetles	. . . .	257
Daddy-long-legs, showing the balancers		262
Common Gnat, grub, pupa, insect emerging and gnat on the wing	. . . .	264
The Hill Ant, *Formica rufa*, and House Ant, *Myrmica molesta*, and their structure	. . . .	271
Ant's Head and Foot, showing the mouth-parts and the leg-combs		273
Section of an Ants' nest, from Figuier		279
Cleared disk of the agricultural ant's nest		295

LIFE AND HER CHILDREN.

CHAPTER I.

*Wisdom and Spirit of the Universe!
Thou Soul, that art the Eternity of Thought!
And giv'st to forms and images a breath
And everlasting motion!*—WORDSWORTH.

I WONDER whether it ever occurs to most people to consider how brimful our world is of life, and what a different place it would be if no living thing had ever been upon it? From the time we are born till we die, there is scarcely a waking moment of our lives in which our eyes do not rest either upon some living thing, or upon things which have once been alive. Even in our rooms, the wood of our furniture and our doors could never have been without the action of life; the paper on our walls, the carpet on our floors, the clothes on our back, the cloth upon the table, are all made of materials which life has produced for us; nay, the very marble of our mantelpiece is the work of once living animals, and is composed of their broken shells. The air we breathe is full of invisible germs of life; nor need we leave the town and go to the country in search of other living beings than man. There is scarcely a street or alley where, if it be neglected for a time, some blade of grass or struggling weed does not make its appearance, pushing its way through chinks in the pavement or the mortar in the wall; no spot from which we cannot see some insect creeping, or flying, or spinning its web, so long as the hand of man does not destroy it.

And when we go into the quiet country, leaving man and his works behind, how actively we find life employed! Covering every inch of the ground with tiny plants, rearing tall trees in the forest, filling the stagnant pools full of eager restless beings; anywhere, everywhere, life is at work. Look at the little water-beetles skimming on the surface of the shady wayside pool, watch the snails feeding on the muddy bank, notice the newts putting their heads above water to take breath, and then remember that, besides these and innumerable other animals visible to the naked eye, the fairy-shrimp and the water-flea, and other minute creatures, are probably darting across the pond, or floating lazily near its surface; while the very scum which is blown in ridges towards one corner of the pool is made up of microscopic animals and plants.

Then, as we pass over plain, and valley, and mountain, we find things creeping innumerable, both small and great; some hidden in the moss or the thick grass, rolled up in the leaves, boring into the stems and trunks of trees, eating their way underground or into even the strongest rock; while others, such as the lion, the tiger, and the elephant, roaming over Africa and India, rule a world of their own where man counts for very little. Even in our own thickly peopled country rabbits multiply by thousands in their burrows, and come to frolic in the dusk of evening when all is still. The field-mice, land and water rats, squirrels, weasels, and badgers, have their houses above and below ground, while countless insects swarm everywhere, testifying to the abundance of life. Not content, moreover, with filling the water and covering the land, this same silent power peoples the atmosphere, where bats, butterflies, bees, and winged insects of all forms, shapes, and colours, fight their way through the ocean òf air; while birds, large and small, sail among its invisible waves.

And when by and by we reach the sea, we find there masses of tangled seaweed, the plants of the salt water, while all along the shores myriads of living creatures are left by the receding tide. In the rocky pools we find active life busily at work. Thousands of acorn-shells, many of them scarcely larger than the head of a good-sized pin, cover the rocks and wave their delicate fringes in search of food. Small crabs scramble along, or swim across the pools, sand-skippers dart through the water, feeding on the delicate green seaweed, which in its turn is covered with minute shells not visible to the naked eye, and yet each containing a living being.

Wherever we go, living creatures are to be found, and even if we sail away over the deep silent ocean and seek what is in its depths, there again we find abundance of life, from the large fish and other monsters which glide noiselessly along, lords of the ocean, down to the jelly-masses floating on the surface, and the banks of rocky coral built by jelly-animals in the midst of the dashing waves. There is no spot on the surface of the earth, in the depths of the ocean, or in the lower currents of the air, which is not filled with life whenever and wherever there is room. The one great law which all living beings obey is to "increase, multiply, and replenish the earth;" and there has been no halting in this work from the day when first into our planet from the bosom of the great Creator was breathed the breath of life,—the invisible mother ever taking shape in her children.

No matter whether there is room for more living forms or not, still they are launched into the world. The little seed, which will be stifled by other plants before it can put forth its leaves, nevertheless thrusts its tiny root into the ground and tries to send a feeble shoot upwards. Thousands and millions of insects are born into the world every moment, which can never live because there is not food enough for all. If there were only one single plant in the whole world to-day, and it produced fifty seeds in a year and could multiply unchecked, its descendants would cover the whole globe in nine years.* But, since other plants prevent it from spreading, thousands and thousands of its seeds and young plants must be formed only to perish. In the same way one pair of birds having four young ones each year, would, if all their children and descendants lived and multiplied, produce *two thousand million* in fifteen years,†

* Huxley. † Wallace.

but since there is not room for them, all but a very few must die.

What can be the use of this terrible overcrowding in our little world? Why does this irresistible living breath go on so madly, urging one little being after another into existence? Would it not be better if only enough were born to have plenty of room and to live comfortably?

Wait a while before you decide, and think what every creature needs to keep it alive. Plants, it is true, can live on water and air, but animals cannot; and if there were not myriads of plants to spare in the world, there would not be enough for food. Then consider again how many animals live upon each other; if worms, snails, and insects, were not over-abundant, how would the birds live? upon what would lions, and tigers, and wolves feed if other animals were not plentiful; while, on the other hand, if a great number of larger animals did not die and decay, what would the flesh-feeding snails, and maggots, and other insects find to eat? And so we see that for this reason alone there is some excuse for the over-abundance of creatures which life thrusts into the world.

But there is something deeper than this to consider. If in a large school every boy had a prize at the end of the half-year, whether he had worked or not, do you think all the boys would work as hard as they do or learn as well? If every man had all he required, and could live comfortably, and bring up his children to enjoy life without working for it, do you think people would take such trouble to learn trades and professions, and to improve themselves so as to be more able than others? Would they work hard day and night to make new inventions, or discover new lands, and found fresh colonies, or be in any way so useful, or learn so much as they do now?

No, it is the struggle for life and the necessity for work which makes people invent, and plan, and improve themselves and things around them. And so it is also with plants and animals. Life has to educate all her children, and she does it by giving the prize of success, health, strength, and enjoyment to those who can best fight the battle of existence, and do their work best in the world.

Every plant and every animal which is born upon the earth has to get its own food and earn its own livelihood, and to protect itself from the attacks of others. Would the spider toil so industriously to spin her web if food came to her without any exertion on her part? Would the caddis worm have learnt to build a tube of sand and shells to protect its soft body, or the oyster to take lime from the sea-water to form a strong shell for its home, if they had no enemies to struggle against, and needed no protection? Would the bird have learnt to build her nest or the beaver his house if there was no need for their industry?

But as it is, since the whole world is teeming with life, and countless numbers of seeds and eggs and young beginnings of creatures are only waiting for the chance to fill any vacant nook or corner, every living thing must learn to do its best and to find the place where it can succeed best and is least likely to be destroyed by others. And so it comes to pass that the whole planet is used to the best advantage, and life teaches her children to get all the good out of it that they can.

If the ocean and the rivers be full, then some must learn to live on the land, and so we have for example sea-snails and land-snails; and whereas the one kind can only breathe by gills in the water, the other breathes air by means of air-chambers, while between these are some marsh-snails of the tropics, which combine both, and can breathe in both water and air. We have large whales sailing as monarchs of the ocean, and walruses and seals fishing in its depths for their food, while all other animals of the mammalian class live on the land.

Then, again, while many creatures love the bright light, others take advantage of the dark corners where room is left for them to live. You can scarcely lift a stone by the seaside without finding some living thing under it, nor turn up a spadeful of earth without disturbing some little creature which is content to find its home and its food in the dark ground. Nay, many animals for whom there is no chance of life on the earth, in the water, or in the air, find a refuge in the bodies of other animals and feed on them.

But in order that all these creatures may live, each in its different way, they must have their own particular tools to work with, and weapons with which to defend themselves. Now all the tools and weapons of an animal grow upon its body. It works and fights with its teeth, its claws, its tail, its sting, or its feelers; or it constructs cunning traps by means of material which it gives out from its own body, like the spider. It hides from its enemies by having a shape or colour like the rocks or the leaves, the grass or the water, which surround it. It provides for its young ones either by getting food for them, or by putting them, even before they come out of the egg, into places where their food is ready for them as soon as they are born.

So that the whole life of an animal depends upon the way in which its body is made; and it will lead quite a different existence according to the kind of tools with which life provides it, and the instincts which a long education has been teaching to its ancestors for ages past. It will have its own peculiar struggles, and difficulties, and successes, and enjoyments, according to the kind of bodily powers which it possesses, and the study of these helps us to understand its manner of existence.

And now, since we live in the world with all these numerous companions, which lead, many of them, such curious lives, trying like ourselves to make the best of their short time here, is it not worth while to learn something about them? May we not gain some useful hints by watching their contrivances, sympathising with their difficulties, and studying their history? And above all, shall we not have something more to love and to care for when we have made acquaintance with some of Life's other children besides ourselves?

The one great difficulty, however, in our way, is how to make acquaintance with such a vast multitude. Most of us have read anecdotes about one animal or another, but this does not give us any clue to the history of the whole animal world ; and without some such clue, the few observations we can make for ourselves are very unsatisfactory. On the other hand, most people will confess that books on zoology, where accounts are given of the structure of different classes of animals, though very necessary, are rather dull, and do not seem to help us much towards understanding and loving these our fellow-creatures.

What we most want to learn is something of the *lives* of the different classes of animals, so that when we see some creature running away from us in the woods, or swimming in a pond, or darting through the air, or creeping on the ground, we may have an idea what its object is in life—how it is enjoying itself, what food it is seeking, or from what enemy it is flying.

And fortunately for us there is an order and arrangement in this immense multitude, and in the same way as we can read and understand the history of the different nations which form the great human family spread over the earth, and can enter into their feelings and their struggles though we cannot know all the people themselves ; so with a little trouble we may learn to picture to ourselves the general life and habits of the different branches of the still greater family of Life, so as to be ready, by and by, to make personal acquaintance with any particular creature if he comes in our way.

This is what we propose to do in the following chapters, and we must first consider what are the chief divisions of our subject, and over what ground we have to travel. It is clear that both plants and animals are the children of Life, and indeed among the simplest living forms it is often difficult to say whether they are plants or animals.

But it is impossible for us to follow out the history of both these great branches or *Kingdoms*, as naturalists call them, so we must reluctantly turn our backs for the present upon the wonderful secrets of plant life, and give ourselves up in this work to the study of animals.

First we meet with those simple forms which manage so cleverly to live without any separate parts with which to do their work. Marvellous little beings these, which live, and move, and multiply in a way quite incomprehensible as yet to us. Next we pass on to the slightly higher forms of the *second* division of life, in which the members have some simple weapons of attack and defence. Here we come first upon the wonderful living sponge, building its numerous canals, which are swept by special scavengers; these form a sort of separate group, hovering between the *first* and *second* division, and from them we go on to the travelling jelly-fish, with their rudiments of eyes and ears, and their benumbing sting, and then to the sea-anemones with their lasso-cells, and to the wondrous coral-builders. Already we are beginning to find that the need of defence causes life to arm her children.

The *third* division is a small, yet most curious one, containing the star-fish with their countless sucker-feet, the sea-urchins with their delicate sharp spines and curious teeth, and the sea-cucumbers with their power of throwing away the inside of their body and growing it afresh. This division goes off in one direction, while the next, or *fourth*, though starting with creatures almost as simple as the coral-builders, takes quite a different line, having for its members mussels and snails, cuttle-fish and oysters, and dividing into two curious groups: the one of the shell-fish with heads, and the other of those without any.

The *fifth* division, starting also in its own line by the side of the third and fourth, includes the creeping worms provided with quite a different set of weapons, and working in their own peculiar fashion, some living in the water, some on the earth, and some in the flesh of other beings, feeding upon their living tissues. An ugly division this, and yet when we come to study it we shall find it full of curious forms showing strange habits and ways.

The *sixth* division is a vast army in itself, with four chief groups all agreeing in their members having jointed feet, and subdivided into smaller groups almost without number. The first group, including the crabs and their companions, live in the water, and their weapons are so varied and numerous that it will be difficult for us even to gain some general idea of them. The other three groups, the centipedes, spiders, and six-legged insects, breathe only in the air. This sixth or jointed-legged division contains more than four-fifths of the whole of the living beings on our globe, and it forms a world of its own, full of interest and wonders. In it we have all the strange facts of metamorphosis, the wondrous contrivances and constructions of insect-life, and at the head of it those clever societies of wasps, bees, and ants, with laws sometimes even nearer to perfection than those of man himself.

Lastly we come to the *seventh* and vast division

of back-boned animals which will claim a separate volume to itself. This division has struggled side by side with the other six till it has won a position in many respects above them all. Nearly all the animals which we know best belong to it,—the fishes, toads, and newts (amphibia), the reptiles, the birds, and the mammalia, including all our four-footed animals, as well as the whales, seals, monkeys, and man himself.

Under these seven divisions then are grouped the whole of the living animals as they are spread over the earth to fight the battle of life. Though in many places the battle is fierce, and each one must fight remorselessly for himself and his little ones, yet the struggle consists chiefly in all the members of the various brigades doing their work in life to the best of their power, so that all, while they live, may lead a healthy, active existence.

The little bird is fighting his battle when he builds his nest and seeks food for his mate and his little ones; and though in doing this he must kill the worm, and may perhaps by and by fall a victim himself to the hungry hawk, yet the worm heeds nothing of its danger till its life comes to an end, and the bird trills his merry song after his breakfast and enjoys his life without thinking of perils to come.

> "While ravening death of slaughter ne'er grows weary,
> Life multiplies the immortal meal as fast.
> All are devourers, all in turn devoured,
> Yet every unit in the uncounted sum
> Of victims has its share of bliss—its pang,
> And but a pang of dissolution: each
> Is happy till its moment comes, and then
> Its first, last suffering, unforeseen, unfear'd,
> Ends with one struggle pain and life for ever."

So life sends her children forth, and it remains for us to learn something of their history. If we could but know it all, and the thousands of different ways in which the beings around us struggle and live, we should be overwhelmed with wonder. Even as it is we may perhaps hope to gain such a glimpse of the labours of this great multitude as may lead us to wish to fight our own battle bravely, and to work, and strive, and bear patiently, if only that we may be worthy to stand at the head of the vast family of Life's children.

CHAPTER II.

LIFE'S SIMPLEST CHILDREN: HOW THEY LIVE, AND MOVE, AND BUILD.

> " The very meanest things are made supreme
> With innate ecstasy.   No grain of sand
> But moves a bright and million-peopled land,
> And hath its Edens and its Eves, I deem.
> For love, though blind himself, a curious eye
> Hath lent me, to behold the heart of things,
> And touched mine ear with power.   Thus, far or nigh,
> Minute or mighty, fixed or free with wings,
> Delight, from many a nameless covert sly,
> Peeps sparkling, and in tones familiar sings.
> LAMAN BLANCHARD.

WHO are Life's simplest children, and where are they to be found? Let us try to answer the second question first, and rubbing the scales from off our eyes, peer into the hidden secrets of nature; and when we have tracked to their home the tiny beginnings of life, we will examine them and try to understand how they live.

How calm, and lovely, and still the sea looks on a warm, sunny, breezeless day of summer, and how happy we can imagine the myriads of creatures to be that float in its waters! We know many of them well, especially those which come close up to the shore. The small fry of the fish, the shrimp and the sand-hopper, the large jelly-fish, and the tiny transparent jelly-bells (see 3', Fig. 22), only to be seen by the keenest eye, as we dip out the water carefully in a glass. Surely these minute jelly-bells with their invisible hanging threads must be some of the simplest and lowest forms of life. Not so, they are really very high up in the world compared with the forms we are seeking.

If, indeed, we come out late some autumn evening when, after the sun has set and the sky is dark, the sea in some sheltered bay appears all covered with a sheet of light, we may see some of the beings of the lowest order of life with the naked eye; for when we dip the liquid fire out in a glass vessel and examine it, we find in it hundreds and thousands of tiny bags of slime giving out the bright specks of light, and these little Noctilucæ, or night-glows (2, Fig. 3), are, as we shall presently see, some of Life's simplest children, although not by any means the most simple of the order.

No; to begin at the very beginning and find the first known attempts at a living being, we must search long and carefully, not merely with our own eyes, but with the microscope. Then we may perhaps be fortunate enough to discover some wondrously small creature like that on the next page, which Professor Haeckel took out of the sunny blue waters of the Mediterranean, near Nice, in 1864. The largest specimen to be found will be smaller than the smallest pin's head, yet when seen under the microscope, this tiny speck appears with outstretched threads, a living animal (see a, Fig. 1), floating in search of food. Examine it how we will, we can find in it no mouth, no stomach, no muscles, no nerves, no parts of any kind. It looks merely like

Fig. 1.

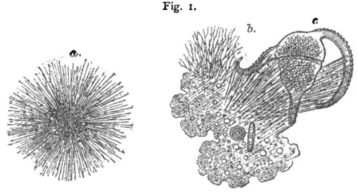

The Thread-slime.*—*Haeckel.*

a, In its natural round shape, immensely magnified; *b*, spreading itself over a small animal, *c* (*Ceratium*), to suck the soft body out of the shell.

a minute drop of gum with fine grains in it, floating in the water, sometimes with its fine threads outstretched, sometimes as a mere drop; and if we take it out and analyse the matter of which it is made, we find that is much the same as a speck of white-of-egg. Is it possible that it can be alive? How can we be sure? In the first place it breathes. If it be kept in a drop of water, it uses up the oxygen in it, and makes the water bad, by breathing into it carbonic acid; then it moves, and, as we shall see presently, can draw in and throw out its fine threads when and where it chooses; again it eats, feeding on the minute jelly-plants in the water, or even on

* Protogenes.

animals higher in the world than itself; and lastly, it grows and increases, for when it is too large to be comfortable it splits in two, and each half goes its way as a living animal.

Let us see how one behaved which Professor Haeckel took out of the sea and kept in a watch-glass under a microscope. When he first looked at it he found that it was drawn up in a lump with a minute animal and a plant-cell in the middle of its slime, and close by it in the water lay a small living animal called a Ceratium (*c*, Fig. 1), which has a hard case or shell. After a while, as he watched, he saw the thread-slime put out its fine threads on all sides (*a*, Fig. 1). Soon the threads on the right side touched the shell of the Ceratium. Here was food, and the body of the *Thread-slime* evidently became aware of it at once, for all the little grains in the slime began to course to and fro, and the threads touching the Ceratium lengthened out and stretched more and more over it, while all those on the other side which had not found any food were drawn in, (*b*, Fig. 1). Six hours later when Dr. Haeckel looked again, to his astonishment the thread-slime had disappeared, but on examining more closely he discovered it completely spread in patches over the shell of the Ceratium. It had drawn its whole body after the pioneering threads and wrapped itself round its prey. Next morning when he looked again, lo! it was back in its original place, and by its side lay the Ceratium shell *quite empty*, together with the skeletons of the other two forms which had been inside the Thread-slime!

This little drop of slime without eyes or ears or

parts of any kind, knew how to find its food; without muscles or limbs it was able to creep over it; without a mouth it could suck out its living body; without a stomach it could digest the food in the midst of its own slime, and throw out the hard parts which it did not want.

This is the history of one of Life's simplest children.

Here is another (Fig. 2), which lives not only in the sea but also in pools and puddles, and in the gutters of our streets and of our house-tops. Anywhere that water lies stagnant these little drops of slime will grow up and make it their home. Sometimes few and far between, sometimes in crowds, so that the whole pond would seem alive if we could see them, they live, and multiply, and die under our very feet. Can anything be less like an animal than this shapeless mass (*a*, Fig. 2)? Yet under a strong microscope it may be seen moving lazily along by putting out a thick slimy finger and then letting all the rest of its body flow after it. When it touches food it flows over it just as the Thread-slime did, and dissolving the soft parts sends out the hard refuse anywhere, it does not matter where, for it has no skin over its body, being merely one general mass of slime.

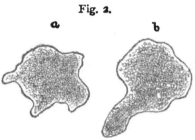

Fig. 2.

a, *b*

The Finger Slime.*—*Haeckel.*
a, At rest. *b*, Feeding on minute slime-plants.

* Protamœba.

And now, before we go on to other forms, let me ask you to pause and think what these little slime-specks tell us about the wonderful powers of Life. Can you guess at all how these creatures do their work? *We* are obliged to have eyes to see our food, nerves and muscles to enable us to feel and grasp it, mouths to eat it, stomachs which secrete a juice in order to dissolve it, and a special pump, the heart, to drive it into the different parts of our body. But in these tiny slime-animals life has nothing better to work with than a mere drop of living matter, which is all alike throughout, so that if you broke it into twenty pieces every piece would be as much a living being as the whole drop. And yet by means of the wonderful gift of life, this slime-drop lives, and breathes, and eats, and increases, shrinks away if you touch it, feels for its food, and moves from place to place, changing its shape to form limbs and feeling-threads, which are lost again as soon as it no longer needs them.

Nor have we yet learnt one-half of the marvels which can be wrought in living specks of slime. For, on further inquiry, we find these simple forms developing two quite different modes of life. In the one case the slime is moulded itself into delicate forms, making creatures with mouths, with suckers, and with delicate lashes to drive the body through the water; while in the other case, remaining a simple drop with delicate threads, it has learned to build a solid covering of the most exquisite delicacy.

To the first class belongs our little Noctiluca, and the forms drawn by its side in Fig. 3. To the second belong the microscopic shells (Fig. 4) which form our chalk. Look at the little wriggling creatures at 1, Fig. 3, small as they look here, they are drawn many thousand times larger than they really are in life, and yet they are much more perfectly formed than either the thread-slime or the finger-slime. They have actually a kind of skin, and do not throw out threads here and there, but are provided with a little

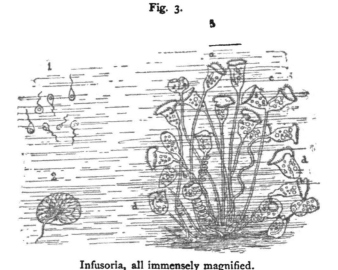

Fig. 3.

Infusoria, all immensely magnified.
1, A group of monads.*   2, The Night-glow.†   3, Bell-flower.‡

whip of slime, which they lash to and fro, and so drive themselves through the water. These microscopic forms called *monads* grow up in water in which flowers have stood for many days till their stalks begin to decay, and in *infusions* of hay or straw, made by pouring hot water upon them and letting it stand; and for this reason the little beings are called *infusoria*. In

* Monas.   † Noctiluca.   ‡ Vorticella.

such impure water, under a powerful microscope you may see them darting along by thousands. But the whip does not only serve them as an oar, it also sends the food they meet with into a tiny opening, one of life's first attempts at a mouth. With a little jerk, when the creature is still or fixed to the bottom, the whip drives still smaller beings than the monad itself into its wide-opened cavity, and there they are digested in a little watery bubble, which may be clearly seen in its body. The Noctiluca or *night-glow* (2, Fig. 3) 's much larger, being often as large as the head of a small pin, and just below the outer rim of its slimy bag the sparks of light are given out. It has been reckoned that there are as many as 30,000 Noctilucæ in one cubic inch of phosphorescent water, and it is almost impossible to grasp the idea of the millions upon millions of these tiny forms which must be floating over a sea which is giving out a glow of liquid fire for miles and miles. And it is only because of this light that we realise that they are there. There are just as many other forms in the water on every side of us, while we dream nothing of this teeming life in the midst of which we live.

We cannot stop here to speak of the *tube-sucker* * and all his relations, which have a mouth at the end of every tube; nor of the beautiful little *bell-flower*,† which may be seen in any pond or in sea-water, with its hanging bells whirling the food in by their little fringe of hairs (*a*, Fig. 3); or shutting up with the food inside, and starting back by curling up their slender stem (*b*); or splitting in two (*c*) and sending off buds (*d, d*), which swim away to form new colonies

---

* Acineta. † *Vorticella nebullifera.*

elsewhere. All these wondrous little beings are some of life's simplest children, and one and all are made of nothing but slime, while yet they live, and move, and seek their daily food.

But all these are naked and homeless, and to a great extent unprotected. Gulped down in thousands and millions by each other, and by other animals, they are defenceless and weak against attacks. It would certainly be better for them if they could have solid shells to cover their soft bodies, and to protect them in many dangers. And so we find that even in this lowest stage of life necessity is the " mother of invention ;" and drops of slime, no higher than the thread-slime (Fig. 1), have learned to build shells around their delicate bodies.

These shell-builders live chiefly in the sea, and there you may find them if you search carefully by the help of a strong magnifying glass in the ooze of oyster-beds, or under the leaves of the delicate green seaweed, or in the muddy sand of the sea-shore. The most common forms will be those shown at *a, e, f,* and *g* in Fig. 4 ; and, though they are so very small, you may if you are fortunate see them clinging by their fine slime-threads to the weeds or the mud.

These animals are, as I have said, simple slime-drops like the *thread-slime,* but they add to the list of wonderful things that such slime can do, for they take out of the sea-water, particle by particle, the lime which is dissolved in it, and build around their soft bodies the solid shell or skeleton in which they live. Nor is this all ; even if they all built the same simple shell, it would be very puzzling to imagine how they do it, but they do much more. They build shells in many different shapes, often with the most beautiful and complicated patterns upon them. All but the simplest shells have several chambers in them, a new one being added as soon as the animal outgrows the last one ; and in the partition between each chamber

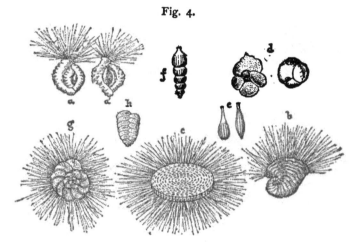

Fig. 4.

a, Miliolite, with a shell of lime. *a′*, The same, with a shell of sand. *b*, Peneropolis. *c*, Orbitolite. *In these shells the animals feed only from the edge of the shell.*

d, Globigerina. *e*, Lagena. *f*, Nodosarina. *g*, Rotalia. *h*, Textularia. *These shells are full of holes, out of which the animal puts threads to feed.*

there is a minute hole through which a thin thread of slime passes into the next chamber, so that the whole body is joined together throughout the shell. On account of these holes these lime-builders have been called *Foraminifera* from *foramen* a hole, *fero.* I bear.*

* This name is now often defined as meaning that the outside of the shell is perforated with holes, but the earlier use of the word as given here is more correct, because it applies equally to the perforated and non-perforated Foraminifera.

Let us now take one of these shells (*a*, Fig. 4), and see how it was built up. The grown animal as he looks when the shell is taken off him is shown in Fig. 5. In the beginning, when he is quite young, he is merely a round drop (1, Fig. 5) with a delicate transparent shell and an opening, out of which he puts his threads of slime. Then as he outgrows this first chamber he draws his slime threads together and forms a bud (2) outside the shell, and round this bud he builds a second chamber out of the end of which he again puts his threads. Then he forms the next bud (3), and goes on thus till he has built a complete shell, generally of seven chambers; and as each new compartment is so placed as to overlap the one before it, the whole when finished has the curious form *a*, Fig. 4, altogether not larger than a millet-seed, from which it takes the name of *Miliola*. These miliolite shells may be found by the help of the microscope in the damp sand of almost any sea-shore, and while some of the shells will be empty, others will still be filled with the dark-yellow animal slime.

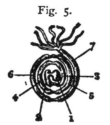

Fig. 5.

The jelly body of the Miliolite, *a*, Fig. 4, showing the buds of slime, 1, 2, 3, etc., round which each chamber is built.— *Carpenter*.

Think of the constant manufacture of such delicate shells as these going on all over the world, and the makers but a drop of slime! And lest you should be inclined to think little of it as a mere mechanical process, the miliolite himself tells us another story, for from time to time we find miliolites with shells made, —not of lime,—but of grains of sand and tiny broken pieces of shell (*a'*, Fig. 4), which the little architect has used to build the walls of his house, when for some reason the ordinary material was deficient. It seems to me that the power of this living drop to choose its own materials is one of the most wonderful facts in the history of life's simplest children.

These miliolites and other Foraminifera when found clinging to sea-weed are easily placed in a salt-water aquarium, and they will then thrust their threads out of the mouth of the shell and crawl on the sides of the glass. Professor Schultze even saw a number of young miliolites born in an aquarium, and this was how it happened. He noticed one day that several of his miliolites had covered the *outside* of their shells with their brown slimy body, and a few days later he could see through the microscope a number of dark-looking specks gradually loosening themselves from this slime.

There were as many as forty of these specks on one shell, and after a time he could distinguish that every speck was a tiny miliolite, having only one chamber (1, Fig. 5) to begin life in, the shell of which was so pale and transparent that he could see the slime within it. As soon as each one shook himself free from the rest of the slime, he put out his threads and crawled away on the glass to get his own living; and now when Professor Schultze examined the shell of the parent miliolite, he found it almost empty. The mother had broken herself up into her little children!

A miliolite builds generally only six or seven chambers, but other forms, such as *c*, Figure 4, build hundreds of separate apartments. This particular form *c*, which is called an Orbitolite, has often as many as fifteen rings, each with its numerous chambers, even when the whole shell is only as large as the head of a small pin; and in ages long gone by, the larger Orbitolites had a far greater number of rings and thousands of chambers in one single shell. The animal builds these in the same way as we have seen the Miliolite do it, only after he has made one round of chambers with a hole in each, he puts out slime-threads at every hole and joins them into a ring with swellings in it, like beads upon a string, and round these he builds the next row of chambers. So he goes on increasing his home till he reaches his full size, and then Professor Parker tells us that the slime of the outer row often breaks up into myriads of young Orbitolites just as the body of the Miliolite did. At the same time these forms can also multiply by merely breaking in half as the naked Finger-slime does, and if by accident a piece of an Orbitolite is broken off it can form a new and complete shell of its own.

If you have now understood how the Orbitolite grows, you will see that the only communication it has with the outer world is through the minute threads which stretch out of the holes of the chambers in the *last* ring (see *c*, Fig. 4), and that the slime in all the middle chambers can get food in no other way than by its passing from the outside right through all the other rings. This is a tedious way of getting food, and we shall find that some of the forms shown in Fig. 4 have escaped from it in a most ingenious way. These forms (*d* to *h*, Fig. 4) have hit upon the plan of keeping their thin threads stretched out like the thread-slime (*a*, Fig. 1) all the time they are laying down their

LIFE'S SIMPLEST CHILDREN.

lime-house. The consequence of this is that wherever a thread has been, there a minute hole like a pin-prick is left in the shell, and while the animal can draw itself quite in out of danger, it can also come out all over the shell and take in food. Here, then, we have another stratagem taught by life to these her infant children. The slime which builds the Globigerina (*a*) or the Rotalia (*g*) is exactly the same as far as we can see as the slime which builds the Miliolite, and yet those drops of slime have learnt a new lesson, and each one as it is born stretches out its fine threads before constructing its shell, thus providing a thousand openings for the entrance of its food in a house not bigger than a grain of sand!

And now it only remains for us to ask how long these wondrous lime-builders have been upon the earth. We ask, and ask in vain, for we have no means of counting the vast ages during which they have lived and built. One of the largest and most complicated forms called the *Nummulite* (from *nummus* a coin, which it resembles), lived and died in such millions before the Alps or the Carpathians had any existence, that whole beds of limestone thousands of feet thick and stretching over hundreds of miles are made entirely of its shells; while the little Globigerina (*d*, Fig. 4) and its friends were living and multiplying in still more dim and distant periods till their shells accumulated into vast beds of chalk.

Fig. 6.

A Nummulite with half the shell broken open, showing the chambers. Life size.

When the ancient Egyptians raised the pyramids

of Egypt, they little dreamed that every inch of the stone they used was made of the shelly palaces of the Nummulite, constructed by little drops of slime with a skill and ingenuity far surpassing their own. As little do most Parisians think now that the limestone of which their houses are built is almost entirely made up of Orbitolite shells. And still less does the country boy as he strolls over the chalk downs of Sussex or Hampshire suspect that the chalk under his feet is largely composed of shells of the Globigerina and the other minute forms shown in Figure 4; yet so it is. These minute slime-builders have been patiently living and building for untold ages, and are doing so still, at the bottom of the Atlantic, where the Globigerina lives in such great numbers that the falling of the shells through the water down to the bottom must be like a constant shower of snow, as is proved by the freshness of those brought up in the dredge.

When a little of the chalky mud was taken up from the bottom at the time when the Atlantic telegraph was laid down, it was found to be almost entirely composed of Globigerina shells, and this led naturalists, who had long known that chalk was formed of shelly matter, to rub down some ordinary chalk and examine it under the microscope, and there again was our little Globigerina, often crushed and worn, but still plainly recognisable. So that, astounding as it may seem, it is nevertheless true that the vast beds of chalk stretching from Ireland to the Crimea, from Sweden to Bordeaux, are in great part formed of the dead shells of these little drops of slime.

We have paused so long over the lime-builders

LIFE'S SIMPLEST CHILDREN.

that we can only glance at those minute specks of slime which build their skeletons of flint instead of lime. These animals are a little higher in the world than the lime-builders, for their body has within it a small bag or capsule, buried in the middle of the slime (see Fig. 7), and in this bag the solid grains lie very thickly, and have sometimes small crystals among them, while in the slime round it there are often little oil-globules floating.

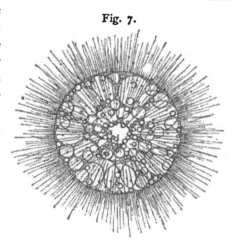

Fig. 7.

The Sun-Slime.*—*Haeckel.
Immensely magnified, its real size being not larger than a mustard seed.

If you dip a glass into the quiet bays of Nice or Messina you may be fortunate enough to bring up one or more of these little sun-slimes, but they are so tiny and transparent that even when the light falls upon them you will only distinguish them as bright specks in the water. Their threads stick out stiff and straight, and for this reason they are all classed under the name *Radiolaria*, or ray-like animals.

Let us look for a moment at Fig. 8, and study the solid skeletons which these Radiolaria build with the flint (or silex) which they find in minute quantities in the water. We saw that the lime-builders construct

* Physematium.

shells into which they can draw back entirely if they are attacked, but the flint-builders seem very careless in this respect, for they have large holes all over their flinty skeletons. But then, on the other hand, notice

Fig. 8.

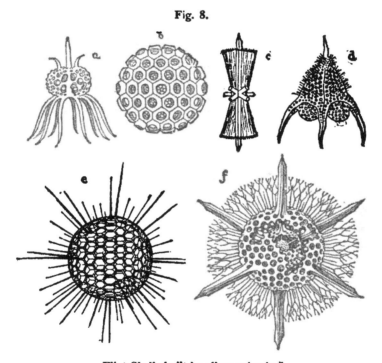

Flint Shells built by slime animals.*

Immensely magnified: the real sizes are from a mere speck to that of one of these letters.

how they send out sharp spikes, which must be uncomfortable for any animal trying to snap at them, although as we have seen (p. 16) the soft thread-

---

* Radiolariæ or Polycistinæ. *a*, Petalospyris; *b*, Ethmosphæra; *c*, Diploconus; *d*, Dictyopodium; *e*, Heliosphæra; *f*, Actinomma.

slime manages to suck their bodies out of the shells. Still these hard spiky outside skeletons must be a great protection to them, and we find every kind of shape devised by these wonderful architects in the construction of their tiny houses, though these are so small as to look like a grain of sand when seen by the naked eye. Perhaps the most wonderful of all is the one shown at *f*, Fig. 8. It is broken open to show the three balls one within another, each kept in its place by rods of flint passing through the whole. This beautiful little shell looks just like the carved balls of the Chinese, yet, instead of being the work of intelligent man, it is built by a mere mass of slime.

We have now learned to know the simplest of all animals ; how they live, and move, and the homes they build. All the forms are not quite equally simple, for some of the higher ones have a solid spot or *nucleus* in the middle of the slime, and sometimes a small watery bubble, as in the Monad or the Bell-flower, which contracts and expands at intervals : and in these forms the outside of the slime is rather thicker than the inside, so that we might say that they are on the road to having a skin, while the shell-builders have a uniform slimy body. But both classes alike belong to that first and lowest branch of the children of life, called by scientific men the *Protozoa* (*protos* first, *zoon* animal) or first animals. The still water everywhere is swarming with them, though we may see and know nothing of them. Yet we owe them something ; for not only do the dead shells of many of them form our solid ground, but those now living purify our waters by feeding upon the living and dead matter in them. These tiny slime animals are the invisible scavengers of the ocean and the pools, and in earning their own living they also work for others. When you look upon a still pond in some quiet country lane, the insects you see swimming about in it, and the plants which cover it, are not the only inhabitants, but on its surface and in its silent depths minute specks of slime are living and working though no eye can see them. Beautiful and wonderful, however, as these forms are, they are yet very low in the scale of life; they live and increase in multitudes, but in multitudes also they die and are devoured. Delicate, and frail, and helpless, they are, as it were, but first attempts at the results which life can accomplish. Let us pass on and see the next step towards higher and, in many ways, more ambitious creatures.

CHAPTER III.

HOW SPONGES LIVE.

> And here were coral bowers,
> And grots of madrepores,
> And banks of sponge, as soft and fair to eye
> As e'er was mossy bed,
> Whereon the wood-nymphs lie,
> With languid limbs in summer's sultry hours.
>
> SOUTHEY.

THERE are certainly very few people, from the little child in the nursery to the artist in his studio, or from the lady in her bedroom to the groom in the stables, who do not handle a sponge almost every day of their lives; and yet, probably, not one in a hundred of these people has ever really looked at the sponge he or she is using, or considered what a curious and beautiful thing it is.

Yet there are at least two things in even the commonest sponges which ought at once to attract attention. If you take a piece of ordinary honey-comb sponge in your hand and look at it, you cannot help being struck by the large holes, few and far between, upon its surface, and the numberless small holes scattered about between them; and on looking carefully down one of the large holes, you will see that it leads to a long tube, into which a number of small tubes open; while, on the other hand, if you try to follow out any of the smaller holes in the same way, you will find that they soon come to an end, and branch out sideways into each other, so as to form an irregular network of short tubes. Lastly, if you cut the sponge open and follow out this network, you will discover that it always ends by leading, sooner or later, into one of the large tubes. What is the reason of this complicated arrangement of holes, all opening into each other, and by whom has it been planned and carried out?

Again, an examination of the material of the sponge will show that it is not a mere structureless mass, but is made up of delicate silk-fibres, woven together into a kind of fine fluffy gauze. By putting a thin slice of the sponge under a microscope, it is possible to distinguish this gossamer tissue very clearly, and to see that it is quite loosely woven; and that it is only because the texture is so fine, and the layers fit so closely one above the other, that, when looked at from above, it appears a solid substance. There is scarcely a more curious object under the microscope than a thin slice of fine sponge, though it is almost impossible in a picture to show its curious nest-like appearance. How has this web been woven so

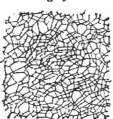

A thin fragment of a bath-sponge seen under the microscope.

delicately? What architect has laid the fibres so skilfully, and formed such a wonderful and intricate structure?

The architect is one of Life's children, whose history we must next consider; for though the sponge was long thought to be a plant, we now know that it is the skeleton or framework of a slime-animal, a little higher than those spoken of in the last chapter. When the sponge which you hold in your hand was alive, growing on the rocks in the warm deep waters of the Grecian Archipelago or the Red Sea, it did not consist merely of the soft fibre you now see, but was covered all over the outside and lined throughout, even along the smallest of its tubes, with a film of slime. This slime, though it appears to be all one mass with specks of solid matter here and there, is really made up of Amœbæ or finger-slime beings (see Fig. 2), and if any little piece is torn off it floats in the water and puts out fingers, exactly as the Amœba does. Nevertheless, in the sponge all these separate cells are not independent creatures, but form the flesh of one single sponge-animal, which lives, breathes, feeds, grows, and gives forth young ones in its ocean home.

At the bottom of the warm seas on the Mediterranean coast or in the Gulf of Mexico these sponge-animals live in wild profusion, sometimes hiding in submarine caverns, sometimes standing boldly on the top of a slab of rock, or often hanging under ledges. Some are round like cups, some branched like trees, some thin and spread out like a fan; while there is scarcely a colour from a brilliant orange to a dull dingy brown, which is not to be seen among them.

The floor on which they grow is often as beautiful as they are themselves, with its covering of tangled seaweeds, among which live the many shelled creatures of the sea, while fish swim hither and thither, and the whole region is teeming with life—

> "Of sea-born kinds, ten thousand thousand tribes,
> Find endless range for pasture and for sport."

Such is the Sponge-kingdom, and the whole colony of sponges of every shape and size flourish like monarchs in their domain. So long as they are alive few can attack them and fewer conquer or destroy them. Only the sponge-fisher diving down into the rich colony disturbs its peace, and tearing the living sponge ruthlessly from its rocky bed, wrings out the living slime, and destroys the animal for the sake of its skeleton.

Every three years this destroyer visits the sponge-colony, for he knows that in spite of his having carried off all the best and richest specimens, this interval is enough for new sponge-animals to have grown up so as to weave large and perfect skeletons.

What secret then has Life taught to the sponge-animal, that while it is still only slime it can grow into such large masses and protect itself so well against the other inhabitants of the sea? We will answer this question by tracing the growth of a sponge from its birth, and reading its history.

If you wish to watch a living sponge yourself, you have only to keep one in a salt-water aquarium, for small sponges are easily found alive on our English coast, though they will not look like those we use. In this description, however, we will imagine that we can visit one of the sponge-colonies in the Mediterranean Sea or the Gulf of Mexico, where the rocks from fifty to a hundred and fifty feet below the surface of the clear blue water are covered with sponges of every size, and shape, and texture.

If we could visit these sponge-beds during the summer or autumn months, and examine carefully the slimy lining of one of the big tubes of a living sponge, we should find that minute bags of slime (1, *a*, Fig. 11) are beginning to appear in it, either scattered through the sponge or collected in heaps. These are sponge-eggs, out of which young sponges are to grow, and in many ways they are very like a hen's egg. Within, as may be seen through their transparent covering, is something which answers to the yelk of an egg, with a solid spot or nucleus in it. This yelk begins soon to divide into two cells, or separate masses of slime, and these again divide into four, these four into eight, and so on till the egg is a globe of small round cells, the beginning of the young sponge. And now a change may be seen to

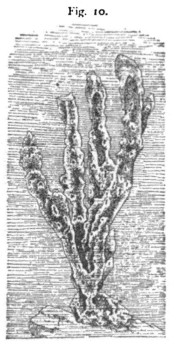

Fig. 10.

A British sponge found at Brighton—life-size.

take place in those cells which lie all round the outside of the rest; each one of them puts forth a minute whip-like lash called a cilium (from *cilium*, an eyelash), so as to form a fringe round the whole body, and then the young sponge, being ready to make its own way in the world, bursts through the skin of the bag, and waving its lashes, swims out an oval-shaped body (2, Fig. 11) into the sea.

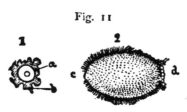

Fig. 11

The birth of the Sponge.—*Adapted from Carter.*

1, Sponge-egg. *a*, The yelk within the envelope, *b*.
2, Young sponge swimming. *c*, Nipple projecting, where a large hole will afterwards form. *d*, Root-cells by which the young sponge afterwards fixes itself to the rock.

Here, you will notice, we have a body, not made as in the simplest slime-animals of a mere piece of slime, but composed of a number of cells, the inner ones round and without lashes, like a group of *Amœba*, while the outer ones, each with his little whip, are like a colony of *monads* (see Fig. 3), surrounding the animal.

By means of these it swims along and feeds; and as it grows, a small nipple, *c*, afterwards to become a hole, appears at the tip, while a group of larger cells (*d*) collect at the hinder end. By means of these cells the little animal attaches itself to the spot where it is to spend the rest of its life, sometimes to a pebble, but generally to the solid rock. Small sponges often fix themselves to living shells, and Dr. Johnstone tells us that he met with a sponge on the back of a crab, which walked about quite unconcerned with its light burden, though it was many times larger than itself.

Having settled, the young sponge now spreads itself out upon the rock, and grows and builds up its fibrous skeleton, while its surface becomes irregular and full of large and small holes, and the true sponge appears.

And now comes the curious part of the story. As the sponge grows larger it is clear that the cells in the middle of its body must be more and more

Fig. 12.

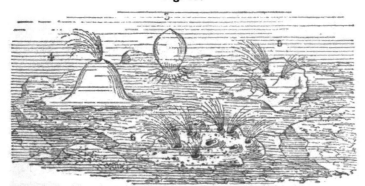

Development of a young English sponge.*—*Adapted from Carter.*

3, The swimming sponge of Fig. 11, which has now fixed itself. 4, The same with water squirting from the hole now formed. 5, The same further developed. 6, The perfect sponge with small holes, where the water enters, and large holes out of which it is squirted.

shut out from the surrounding water out of which food can be taken, and yet these cells want feeding as much as those outside. In order to bring this about, the sponge-animal, instead of growing up as a solid mass of slime-cells, arranges the silky fibres of its skeleton in such a manner as to leave a number of small canals or passages throughout its body, and these open, as we have seen, sooner or later, into large canals

* *Halichondria simulans.*

or main thoroughfares, while the slimy sponge-body is spread out as a thin film along them all. In this way it is possible for the sea-water to reach right throughout the whole body of the sponge along the various canals. But if this water only lay still from day to day, no fresh food could be brought, and the whole would become stagnant and bad. The animal cannot feed or even breathe unless a constant fresh supply of water, full of oxygen and living beings, is driven through the canals.

How is this to be done?

At first sight it seems as if the young sponge were behaving very foolishly in this matter, for no sooner has it settled down than it draws in all the whip-like hairs outside its body which we should have thought would be useful for driving in food, and becomes a mass of smooth slime-cells with large and small holes scattered here and there. Still, as the water goes on pouring out at the big holes (see Fig. 12), it is clear that it must be going in somewhere; and on cutting open the living sponge and watching it at work the secret appears. Here and there throughout the narrow canals of the skeleton are to be found little chambers, like two saucers face to face (1, Fig. 13), and in these are arranged in rows a number of whip-like cells, exactly like those which were before outside the sponge. It is the whips in these cells which do the work required. Waving ceaselessly to and fro, they drive the water before them always in one direction, so that it is drawn in at the small holes ($a\ a$, Fig. 13) and driven out at the large ones ($b\ b$). By means of this wonderful contrivance fresh sea-water full of oxygen and living plants and animals is always pouring along the small canals, bringing air and food to each cell along the road, while the bad water out of which the slime has taken all the oxygen, and into which it has thrown the hard parts and refuse of its food, is driven out at the large holes, carrying away with it all that is hurtful and useless.

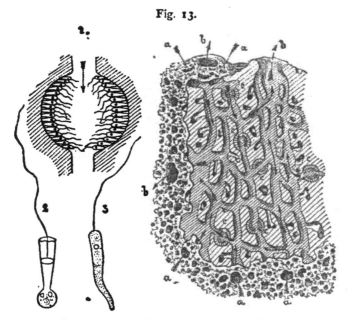

Fig. 13.

Section of an ordinary bath sponge.—*Murie.*

a a a, Small holes where the water enters. *b b*, Large holes where t flows out. *c c*, Chambers with whip-cells which drive the water on. 1, A chamber enlarged showing the cells. 2, 3, Different forms of whip-cells.

And now we can understand why the sponge, though a mere slime-animal, is classed as the pioneer of the second division of living animals; because in it quite a new plan of structure has begun. Starting from one egg, the whole sponge is one single individual; yet, when grown up it is not a mere mass all doing the same work, as in the simplest animals, for it has learnt the secret of division of labour; and while one set of cells, those forming the smooth slime, are busy taking in food, the other and whip-like cells are foraging for this same food and sweeping away the refuse; and, between these two, a special layer of smooth cells is employed in building up the skeleton which supports the whole body.

If we knew only the grown-up sponge, we might look upon it as a society of two kinds of slime-animals living together and building a common house. But when we consider that each whole sponge comes from a single egg, growing and dividing like one of the eggs of the higher animals, and that any piece of a sponge-animal is able to settle and grow up into a perfect sponge with the two kinds of cells, we see that these animals have made a great step never again to be forgotten by the children of Life. They have learned to form in one body two kinds of cells with different duties, which, by their mutual labour, carry on in one being the work of life.

We are now, I hope, able to picture to ourselves the sponge growing upon the rocks in deep water, or sometimes in shallow pools, or between the tide-marks, looking like a smooth mass of slime of different shapes, with holes invariably open as long as it is under water, but closed (as we shall find on the English shores) if the sponge is by chance left high and dry by the tide. We can imagine to ourselves the small fountains of water spouting from the larger openings, and carrying off the refuse from the inside of the sponge, and we can fancy we see the small chambers buried in the canals with their active inmates lashing the water onwards in its course through the whole mass.

But we have yet to consider the *skeleton* of the living animal, and why so much time and labour should be spent in forming it. There are two reasons why a solid framework is useful to the sponge-animal. First, it supports the large mass of soft slime, and enables it to spread itself out in thin layers, so as to touch the water in the canals; and, secondly, it protects it from enemies.

There are a few sponges made entirely of slime, the canals and thoroughfares being in the slime itself; and in these, when the animal dies and decays, nothing solid is left behind. But such sponges have probably become degraded and have lost their skeleton, and they are clearly under a disadvantage, for the walls of slime are forced to be much thicker, and food cannot reach them so easily; and besides this, when we remember how many sea-animals feed on living slime, we cannot but see that these sponges offer a very tempting feast. Comparatively large animals, such as shrimps and fish, will take big mouthfuls out of them, while the water-fleas and smaller sea-worms which are carried through their canals, are quite as ready to eat the slime as the slime is to eat them. But if the sponge can offer a very tough and unpalatable mouthful, or can prick its enemies' mouths with a sharp point, they will not be so ready to take a second bite; and so it comes to pass that we find in sponges some of the most curious weapons imaginable.

The sponges we use are by no means the first attempts at sponge-skeletons; on the contrary, they represent the highest art in sponge-building. The simplest kind of sponges build their skeletons of lime and flint, as did the earlier slime-animals. Fig. 14 is a picture of a lime-sponge. Here the outer layer of sponge-flesh has taken in lime and built up with it a number of little pointed spikes or *spicules*, which lie buried in the slime. The rest of the sponge is composed entirely of the sponge-animal, the outer cells being smooth and the inner ones whip-like, so that water and food are drawn in at the small holes in the sides, while the refuse is driven out at the large hole in the top.

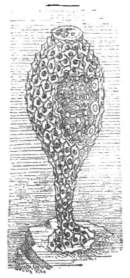

Fig. 14.

Sponge with lime spicules forming the skeleton over the living flesh.—*Haeckel.**

Now suppose that a fish attacks this sponge, instead of a mouthful of soft slime he will bite upon a number of minute sharp points which he will carry away sticking to the soft lining of his mouth, and the next time he sees such a sponge growing, he will hesitate before touching it. In some sponges these lime-thorns are so arranged that they lie flat against the sponge when it is still, but form a complete hedge of spikes round the holes when it is taking in water, showing that it is not only against the fish that it is protecting itself

* *Ascetta primordialis.*

HOW SPONGES LIVE.

but against the smaller but dangerous animals, which might be washed into it. In another sponge the spicules point towards the mouth at the top, so that any creature which has got in can be easily thrown out but one trying to get in would be spiked directly.

Lime-sponges are to be found in most parts of the world, and some of them are very beautiful from the arrangement of their spicules. But these look, after all, like mere rough attempts at spike-building when compared with the wonderful spicules which are made by the flint-building sponges.

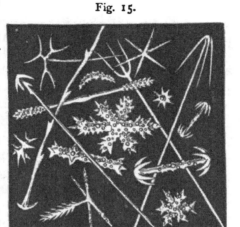

Fig. 15.

Spicules of flint found in the flesh of flint-sponges. Real size a mere speck, almost invisible to the naked eye.

Fig. 15 shows only a very few of the forms of flint spicules which are known. They look, under the microscope, as if the sponge-animal were an artist trying how many curious patterns he could invent; and yet Dr. Bowerbank has shown that each of these shapes has some special use, either in keeping out enemies, in supporting the sponge, or in spiking and entangling the smaller animals which form the food of the sponge-animal. Often as many as from three to seven different shapes may be found in one single sponge, forming by their combinations intricate and beautiful patterns. Yet each one of these spicules, perfect and complete in form as it is, is so small as to be barely visible as a speck to the naked eye, and so transparent that when mounted on glass for the microscope it is impossible to detect even a group of them without a lens.

In Fig. 16 may be seen three kinds in their natural position in the flesh of the sponge, the large ones binding the sponge together, and the small feathery and anchor-shaped spicules protecting the flesh; and small as these last appear, yet they are even now magnified 100 times. Lastly, in the higher flint-building sponges the architect gets beyond mere separate spicules, or binds them together so skilfully with fine, transparent flint threads that they form a network of wondrous beauty. Looking at the marvellously delicate Venus' basket (Fig. 17) which grows in the seas near the Philippine Islands, it is almost impossible to persuade ourselves that the flint-lace of which it is made has been constructed by an animal with no eyes to see the beautiful pattern it was weaving, and no machinery in its body with which to direct the web; and that out of mere slime cells has arisen a fairy structure such as the most skilled human artist might try in vain to rival! These sponges live chiefly in very deep water. In one of them, called the glass-rope sponge, the animal is

Fig. 16.

A piece of a flint-sponge with the *sarcode* or flesh, magnified 100 times.
—*From life.*

HOW SPONGES LIVE. 47

anchored to the bottom by long flint threads, often several feet long, looking like the finest spun glass.

And now we find the sponge-animal advancing yet a step farther, and beginning no longer to build entirely with lime and flint taken from the water, but to manufacture its own material. We all know that the spider spins its web of threads of gum formed in its body, and that the silk of the silkworm is made in the same manner, and now we have to learn that the sponge‑animal with its simple slime cells can do this too! For all the sponges which we use are made of fine fibres, which prove, when examined, to differ very little chemically from the silk of the silk‑worm. These fibres have been secreted by the slime‑animal out of its food, and by crossing and re-crossing them in all directions it forms the soft elastic skeleton of the toilet sponge. Yet they are not woven carelessly or without purpose, for we have seen that they

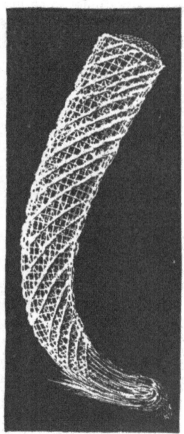

Fig. 17.

Venus' Basket.* The skeleton of a flint-sponge.

* *Euplectella speciosa.*

are so arranged as to build up the small canals and the large tubes in their right positions; and though all may look confused to us, yet there is no part which the water cannot reach in its passage through the sponge.

At first in the coarser sponges the fibre is thick and loosely woven, and though it is tough and almost impossible to bite or digest, yet it leaves such large openings as to afford but a poor protection. In these sponges flint spicules are still built in with the fibre, scattered about in all directions: and, because of the sharpness of the spicules, their skeletons are of very little use to us. But little by little, in sponges of a finer web, in which the tough silky fibres are so closely matted together as to repel all intruders, we find the sponge-animal beginning to neglect the formation of spicules, and contenting itself with building in fragments of sand, making those gritty sponges so disagreeable to handle. And by and by it ceases even to do this, and in the fine soft Turkey sponge we find the holes so small that no enemy large enough to do harm could enter, while the densely woven fibres offer a most unpalatable and indigestible morsel to any creature which might have

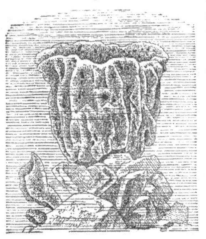

Fig. 18.

Cup-sponge growing in the sea. Real size about a foot high.—*From Figuier.*

the strength to tear it away; and these, needing no further protection, are made entirely of soft fibre.

Here we must leave the history of sponges and their lives. We have left much unsaid, for to tell how sponges may increase by dividing or by budding, as well as by eggs, would have taken us too far into detail; neither could we give space to trace the wonderful way in which the various spicules are used as weapons of defence; and for special examples of the different kinds of sponges you must consult works on natural history. We have had one chief object in view, namely, to see how Life in this new form has advanced beyond the earliest slime-animals. The sponge, with its two forms of cells and its division of labour, stands already far above the microscopic beings of our last chapter. Rooted to the rocks, and large enough to invite the attacks of enemies, it has yet learnt to protect itself by wonderful structures, to distribute its food throughout a large body, and last, but not least, no longer to form its skeleton merely of flint or lime, but to manufacture in its own body the material with which it builds.

It has indeed succeeded so well that Dr. Bowerbank, one of the best authorities on sponge life, came to the conclusion that sponges are able to escape almost entirely, during their lifetime, from becoming the food of other animals. It is only after their death that their slime serves to nourish myriads of minute creatures, and then the wonderful rapidity with which the living matter is devoured, is quite enough to prove to us how well the living sponge must have used its weapons to protect itself, while still it was one of Life's living children.

CHAPTER IV.

THE LASSO-THROWERS OF THE PONDS AND OCEANS.

" Transparent forms too fine for mortal sight,
Their fluid bodies half dissolved in light."

Millions on millions thus, from age to age,
With simplest skill, and toil unwearyable,
No moment and no movement unimproved,
Laid line on line, on terrace, terrace spread,
To swell the heightening, brightening, gradual mound,
By marvellous structure climbing tow'rds the day,
Each wrought alone, yet all together wrought
Unconscious, not unworthy instruments
By which a hand invisible was rearing
A new creation in the secret deep.
MONTGOMERY.

IF among all the children of life we wished to choose out the most brilliant, graceful, and sylph-like creatures whose histories are more like fairy poems than sober reality, we could scarcely do better than select those which we are now going to study under the name of the "lasso-throwers," and strange as this name may appear, I hope to show that it is not too fanciful to be accurate.

Every one knows that the long cord or thong called the lasso is the peculiar weapon of the South American hunter. From his earliest childhood the

THE LASSO-THROWERS.

young Gaucho learns to play with the lasso, and almost as soon as he can walk, amuses himself by catching young birds and other animals round his father's hut, throwing out the long lash so skilfully that the noose falls over their bodies and brings them to his feet. As soon as he can ride he carries the sport farther, galloping wildly over the plains swinging the cord round his head and letting fly at the ostriches, the wild cattle, and horses, or when he is a man, even at the jaguar or the puma. Such is the lasso as man uses it, consisting of a long cord or thong thrown with exquisite skill.

Now among animals, as we have already seen, any weapons they are to use must be such as grow upon the body, and we should little suspect that a simple jelly-animal could be provided with a lasso ready grown within its flesh. Yet so it is. In that division of life's children, standing in rank just above the sponges, we find a weapon of this kind as simple, as deadly, and far more wonderful in its action than the lasso of the American hunter.

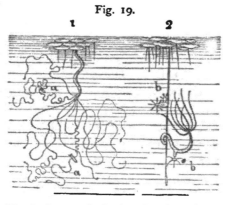

Fig. 19.

The fresh-water hydra hanging from duckweed in a pond.

1. The long-armed hydra* feeding. *a a* small animals caught in its arms. 2. Short-armed hydra† throwing off young hydra-buds, *b b.*

In almost any wayside pond in England it is pos-

* *Hydra fusca.* † *Hydra viridis.*

sible to find, either hanging from under the leaves of the common duckweed, or clinging to pieces of floating stick, or rooted to stones at the bottom of the pond, a little greenish being (Fig. 19), about a quarter of an inch long, looking like a tube with a circlet of feelers at the end, which are waving in the water. This creature is the common pond hydra, and it is in fact nothing more than a tube or sac, with a sucker at one end to hold on with, and a number of jelly-arms or *tentacles* at the other, which serve to catch its food, and to tuck it into the sac, where it is digested. The walls of the sac are firm and muscular, and the creature can stretch itself out, or draw back at will, can move along slowly by means of its sucker, and even float upon the water; but the most remarkable thing about it, and the one we wish to study now, is the power which it has of overcoming animals stronger and more active than itself.

Groping about with its flexible arms, which are covered with fine jelly-hairs by which it seems to feel, it touches perhaps a water-flea, a water-worm, or even a tiny newly-born fish, passing by in the water. Instantly it twists its arms round whatever it finds, and though its prey may struggle vigorously while the hydra remains almost still, yet little by little the struggles cease, and the victim is drawn into the fatal sac.

Now, why is this? It is because those fine tender feelers of the hydra are full of lassos, which it can use with as good effect as any skilled hunter. Although to the naked eye each tentacle looks but little more than a fine hair, yet, when examined under a strong microscope, it is seen to be crowded

THE LASSO-THROWERS.

with hundreds of clear transparent cells (1, Fig. 20), so small that 200 of the largest of them would lie side by side in an inch, while many are not more than $\frac{1}{1500}$th of an inch long, and each of these cells contains a formidable weapon. Within the cell, lying bathed in a poisonous fluid, is coiled a long delicate thread, barbed at the base (2, 3, Fig. 20), and this thread may well be called a lasso, since it is always many, and often from twenty to forty times as long as the cell itself, and only waits for the bag to burst to fling itself out to its full length.

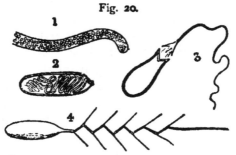

Lasso-cells of the Hydra and Sea-Anemone.

1. Piece of one of the hydra's arms, showing the cells crowded in it. 2. One of the cells. 3. The same cell after bursting open. 4. Lasso-cell of an anemone.

Now looking at the hydra, Fig. 19, picture to yourself that each of its delicate thread-like tentacles is crowded with hundreds of these lasso-cells, only waiting the word of command to discharge their weapons. By and by the two worms (*a a*) come within reach, and rub against the tentacles, instantly every cell that is touched bursts open, and with a spring its lasso is set free and shoots out, piercing through the skin of the worm.

And now we can see where the hydra's strength lies. He has no need to struggle, for his victim is pierced by a number of darts, and the poisonous fluid from the cells is pouring into him. And there is great reason why the hydra should take it so quietly; he does not wish to waste his lassos, for a cell once burst cannot be used again, and he will have to grow a new one for each one that he exhausts. So he waits patiently for the spell to work, and does not hug his victim too close until he is half conquered, and then he draws him gently in.

So the hydra lives and catches its food without needing to move far from the place of its birth. All the summer through it puts out buds (see *b b*, Fig. 19) from its side, and these buds, as soon as their tentacles are grown, drop from their parent and settle in life for themselves, so that any pond may contain hundreds of them; and when the winter comes, and before they all die, an egg appears near the base of the tubes of those which are then living, and these eggs lie till next spring, when they are hatched, and produce a new generation of hydras.

This is the simplest lasso-thrower, and I think you will allow that his lasso is both wonderful and deadly, so that, though these hydras are the only lasso-throwers to be found in fresh water,[*] it is easy to understand that his relations in the wide ocean should have made good use of the new weapon with which life has provided them, and secured homes and resting-places throughout the whole world of waters, and under all kinds of strange shapes and forms.

From the North Sea to the Tropics, from the pools on the shore at low tide to the depths of the wide ocean, we meet everywhere with this division of "lasso-throwers." Now in the shape of large jelly-fish

---

[*] Since the above was written, a fresh-water Medusa has been found in the tank of the Victoria Regia in the Botanical Gardens, Regent's Park.

covering the sea for miles and miles, so that a ship may sail through them during many days, the sailors watching their transparent domes by day, and being illuminated by the light of their phosphorescence by night. Now as tiny jelly-bells, floating like glistening specks by millions in some quiet bay, and breaking into light as they are dashed upon the beach. Or again in the form of horny animal-trees often two or three feet in height, waving their gracefully arched branches over the rocks in the deep water, or creeping like delicate threads over shells and stones and seaweed on the shore, where they are often mistaken for plants.

There is scarcely a nook or cranny in the bed of the ocean where some of these tree-like forms are not to be found, associated with the beautiful sea-anemones, with their brilliant colours of emerald green, crimson, glowing purple, and vivid orange, which belong to this same division, as does also the living coral nestling in the bosom of the warm Mediterranean Sea, or struggling boldly against the waves of the Pacific, as branch after branch is added to its stem by the constant labours of the tiny jelly-polyps spreading their gaily coloured tentacles out of every cup of the coral tree.

All these beautiful creatures are "lasso-throwers." Scientific men call them *Cœlenterata* or "hollow-bodied animals," because of the large cavity within their bodies, and divide them into *Hydrozoa* (water-animals) and *Actinozoa* (ray-like animals, such as the anemone), but for us it is sufficient to know that, with very few exceptions, they all seize their prey by means of the lasso, and we can pass on to learn something of how they pass their lives.

It is scarcely possible to collect seaweed on any coast without finding upon it what look like minute plants (Fig. 21) with frequent joints. Some of these, which are formed of chalk, are true plants, but others, which are yellowish and horny, are no less certainly animals, and you may soon detect these by means of a magnifying glass, for they will bear at each joint a little cup (*c c*), and if you could watch these cups

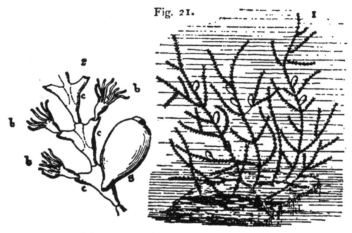

The Sea-Oak.*—*Adapted from Hincks.*

1. The animal-tree growing of the natural size. 2. A piece of one branch enlarged, showing the animal *b* stretching out of the horny cup *c*, and one of the egg-sacs *s*.

when the creature is alive, you would see out of each one from 12 to 16 transparent tentacles (*b b*) sweeping round in search of food.

This tree-like stem is in fact the home of a hydra of the sea. The creature itself is like the pond hydra, only that its buds do not fall off but continue to live

* *Sertularia plumula.*

all together, each enclosed in a cup made of a peculiar substance called *chitin*, which is nearly allied to horn, and which also forms the skin of insects. The whole stem is only one individual, for a fine living thread passes down through the bottom of each cup and meets all the others within the stem, so that the food digested in each tiny stomach goes to feed the whole animal.

Here then we have hundreds of tiny lasso-throwers acting as mouths and stomachs to one *Sertularia*, as this specimen is called. Each mouth or *polypite* is so small as scarcely to be seen even as a speck by the naked eye, yet it has sixteen tiny arms, and each arm is crowded with lasso-cells!

And now in the summer months, between May and September, small round bags (*S*, 2, Fig. 21) appear scattered along the branches of this animal-tree, and each one of these is full of eggs; and by and by, when the eggs are hatched, young sertularians swim out as little round jelly bodies, and settling down on some stone or seaweed grow up into new stems of lasso-throwers.

It is scarcely possible to conceive the number of minute beings which are feeding in this way at the bottom of the sea. This particular sertularia or *sea-oak coralline* (Fig. 21) covers the seaweed of our coasts with miniature animal forests, and yet it is one of the smaller kinds, sometimes not more than half-an-inch high. Others grow on shells forming a fleecy covering which looks only like a little white moss, but which is really a group of living animals.

Every child must be familiar with a kind of rough crust frequently to be seen outside old shells, but probably few have ever thought that this is often the remains of the home of a lasso-thrower, any more than they would connect the tube-like branches on the seaweed with a living animal.

Try for one moment to picture to yourself some quiet spot in the ocean-bed, where the whole floor is carpeted with such forms, and every shell and seaweed carries some hundreds of tiny beings all stretching out their waving tentacles and flinging out their miniature lassos to strike their prey. You would see here and there among them tall, graceful, animal trees, such as the sea-fir,* which often grows up into a brown upright tree more than three feet high, with branches bearing as many as a hundred thousand cups, each with its pure white polypite stretching out, and looking wonderfully delicate against the dark stem; while side by side with it may be standing the tube-hydra,† which has single yellow pipes, out of each of which a brilliant scarlet creature is waving its graceful tentacles. All this life is active and busy, and yet it is all made up of beings so insignificant to us that we have hardly any idea of their existence.

And while you were watching these thousands of tiny arms you might perhaps witness a strange sight if your eyes were sharp enough to see it. From an animal tree very like the sertularia (1, Fig. 22), except that its horny cups are borne upon stalks, you might see escaping some little beings looking like green shining bubbles, and these, if seen under the microscope, turn out to be the most beautiful fairy jelly-bells (3′, Fig. 22), like pure crystal domes

* *Sertularia cupressina.* † *Tubularia indivisa.*

THE LASSO-THROWERS.

swimming gaily along in the water by driving in and out the jelly-veil (*v*, 3) spread across their rim.

Anything to exceed the delicacy and beauty of these tiny jelly globes can scarcely be imagined, and

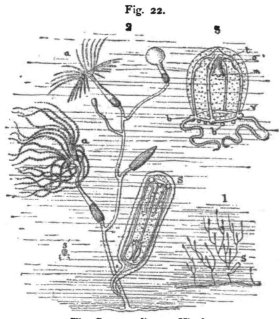

Fig. 22.

The Campanulina.—*Hincks.*

1. Natural size of the animal tree. *s*, Natural size of the sac containing the jelly-bell. 3'. Natural size of free jelly-bell floating in the water.

2. A piece of the fixed animal-tree magnified. *a a*, Animal feeding. *s*, Sac with the jelly-bell upside down inside it.

3. Free jelly-bell magnified. *v*, The veil across the bell. *t*, Feeding tube of the animal. *m*, The mouth. *o*, The ovary in the canals of the bell. *b b*, Coloured spots in the rim.

as they are easily bred in a salt-water aquarium, all their life-history may be carefully studied. Minute as they are, the pulsation of the bell as they propel themselves along may be distinctly seen with the naked eye; and when put in a drop of water under the microscope, all the different parts of the body, as shown at 3, Fig. 22, can be clearly made out. For this bell is a true and delicately organised living being. It is a new instrument which Life has invented for carrying the eggs of the animal-tree far away over the sea. While the mouths, $a\ a$, are busily catching food for the whole animal by their lassos, there has been growing on part of the stem a bag (S), in which this little bell has been formed, and when it is ready to start on its journey the bag opens at the tip and the bell struggles out. How gracefully it now drives itself along by shooting water in and out of the hole in its thin veil as it contracts and expands its rim, and from the water thus driven in, its mouth (m) takes the minute living beings, and digesting them in its tube t, sends the nourishment down the canals to the rim, and so over the whole bell; while in the little bags o in the canals it forms and carries the eggs to be dropped down on some distant spot to grow up into a new animal-tree.

Thus this minute bell is a living, active creature with all the necessary parts for swimming and feeding, and also for forming eggs to give birth to young ones by and by. Its whole body is crowded with lasso-cells, though it does not seem greatly to need them, and what is much more interesting, in many cases it even bears on its rim the first attempts at eyes and ears.

Often the passage of these tiny bells through the water can only be traced by some bright spots like coloured gems set in its rim (*b b*, 3, Fig. 22). Blue, scarlet, orange, all the most vivid colours seem chosen to give them brilliancy, and inside the spots are in some cases to be found little grains of lime which roll to and fro and probably form the simplest hearing apparatus in nature, while some crystals which refract light are the first beginnings of eyes.*

Is it too much to say that these minute jelly-bells are fearfully and wonderfully made, and that our imagination sinks appalled when we have to believe that such complex beings have sprung from the tiny buds on the animal-tree (1, Fig. 22)?

In the early summer the sea is full of these little bells rising like constant bubbles from the animal-forest below. Some are mere microscopic specks, others as large as thimbles, while some look like glass cups floating in the sea. They are all more or less tinted with lovely and delicate colours, and though an unpractised eye cannot distinguish them, yet they may be caught in a fine muslin net swept through the water and examined under a microscope, or in an aquarium; while on a calm evening, when the sea breaks in ripples on the sand, their presence is betrayed by the glow of phosphorescence fringing the shore.

"Figured by hand Divine, there's not a gem,
Wrought by man's art, to be compared with them.
Soft, brilliant, tender, through the wave they glow,
And make the moonbeams brighter where they flow."

And now we will rest our eyes from straining to see the microscopic lasso-throwers and turn to some

* In the higher forms of Medusæ or jelly-fish the presence of nerves has now been clearly proved by Hertwig, Romanes, Schäfer, and others.

of the large jelly-fish of the sea; for after watching the floating-bells we cannot doubt that those enormous jelly masses which we see sailing along in the ocean are their near relations. Indeed, those who swim and bathe in the sea can testify feelingly to the power of the poisonous lassos, for to be stung by a jelly-fish is no slight matter, and this sting is given by the lasso-cells.

Though jelly-fish, however, are uncomfortable to meet in the water, they are most interesting to watch from a boat, or the head of a pier, as they move along dome foremost, with a regular movement, as if by clockwork. We scarcely realise how large they are, till coming close to them we lay an oar over them, and find perhaps that the dome measures a foot and often two or three feet across, while their tentacles stretch from the head to far beyond the stern of an ordinary boat. From spring to late autumn they may be seen when the weather is calm, sailing on the water, not by means of a veil like the jelly-bells, for they have none, but by the movements of their huge umbrella, which they contract when storms arise, and so sink down into the depths. What is the history of these huge soft masses?

First we must notice how very little solid matter life has to use in building up their bulky forms, for when a jelly-fish of four or five pounds' weight is cast on shore and dried up by the sun, a film weighing a few grains is all that remains; the rest was all water! Yet the creature is wonderfully made. Take for instance the jelly-fish shown in Fig. 23. Its tawny, yellowish umbrella is full of canals carrying the nourishment over the whole animal. In its margin

THE LASSO-THROWERS.

rudimentary eyes and ears are covered with a delicate hood to shield them from harm. Powerful muscles contract and expand the rim of the umbrella, guided by nerves lately discovered in these animals, while the mouth of the hanging stomach (itself hidden under the umbrella) has long, tawny lips which trail behind it like ribbon sea-weed, and are most formidable weapons, for they are crowded with powerful and poisonous lasso-cells. A creature which this jelly-fish has once seized in its lips must die, for even if it can get loose from the strong grasp, the poison works and it soon floats dead on the water.

Fig. 23.

A jelly-fish,* whose life history is given in Fig. 24.

Shrimps, barnacles, small fish of all kinds, and even the strong cuttle-fish and other larger

* *Chrysoara hysocella.*

animals of the sea, are devoured by this ferocious lasso-thrower as he moves lazily through the water expanding and contracting the rim of his dome; and if it were not that he and his fellows are the chief food of whales and porpoises, they would commit terrible havoc in the ocean, as they travel in shoals of thousands together.

And now at certain seasons of the year, when at night the sea glows with their phosphorescence,* some of these large wanderers drop from under their huge umbrella something which looks like a shower of dust. This shower is composed of a number of minute jelly-bodies (*a*, Fig. 24) swimming by means of lashes or *cilia*, and something like those which come from a sponge (see p. 38). They have been hatched from eggs within the umbrella of the jelly-fish, and are setting off into life for themselves. After a few days four curious knobs (*b*, Fig. 24) begin to appear upon them, and these increase every day till at last the swimming animal settles down on a rock and becomes a small hydra feeding peacefully upon minute sea-animals by means of its tender threads.

This then is the young of our jelly-fish, a common hydra like that of the pond! Moreover, this young hydra seems to forget all about its wandering parentage, and often goes on for several years budding into other hydras, and living as though it had never had anything to do with a jelly-fish.

But at last one day a change comes over some of

---

* The phosphorescence is due to a glutinous fluid exuded from the umbrella. This fluid, when squeezed from a large jelly-fish into twenty-seven ounces of cows' milk, made it so phosphorescent that a letter could be read by the light at a distance of three feet.

THE LASSO-THROWERS.

the hydras of the colony, which may be great-great-grand-buds of the hydra which settled down.

They lengthen out and their bodies divide into rings (*c* and *d*, Fig. 24), and as these rings grow deeper and deeper the tentacles fall away from the

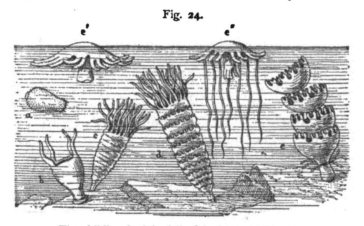

Fig. 24.

The childhood of the jelly-fish shown in Fig. 23.

a, Swimming jelly-body. *b*, The same fixed to a rock, which may go on budding like a hydra and giving off forms like itself for many years. *c*, The hydra beginning to divide into rings. *d*, Rings becoming more perfect. *e*, Rings breaking off from the hydra. *e'*, One ring which has turned over and begun to grow into a jelly-fish. *e"*, The same developing into the perfect jelly-fish as in Fig. 23.

top and begin to grow out below (*e*), and at last one of the rings drops off from the top, a complete saucer (*e'*), and turning over so that the domed part is uppermost, begins to contract and expand its rim, and sails away a minute jelly-fish! Other rings follow in its path, and the descendant of the fixed hydra has again become a group of wandering lasso-throwers.

And now the floating domes begin to grow rapidly; in each one the umbrella thickens, the stomach with its huge lips begins to lengthen and expand, the eye-spots develop under their hooded covering, the tentacles sweep out into the sea, and the shoal of terrible monster jelly-fish is abroad again

> . . . . . "all in motion
> Far away upon the ocean,
> Going for the sake of going,
> Wheresoever waves are flowing,
> Wheresoever winds are blowing."

And here we must leave them. The history of all jelly-fish is not exactly alike, for they do not all go through the strange transformations just described. The beautiful purple Portuguese man-of-war, with its rose-tinted jelly-sail, is born a wanderer like its parent, and so are also the lovely "Hanging-Bells,"* which have from ten to twelve, and even sometimes as many as sixty, clear, transparent bells hanging from their stalk, like blossoms on a flower, while a clear bubble shining like quicksilver serves as their float. These and many others have each their special history for those who care to study them, and even this brief glance at the wandering lasso-throwers will surely lead us to look with more interest on the shapeless dying lump of jelly on the sea-shore, now that we know it to have been an active living animal with powerful weapons, sensitive nerves, and jewelled eyes.

After following the free adventurous life of a travelled jelly-fish, it seems almost like visiting some quiet little country village, to turn to the dreamy sea-anemones, living from day to day in their rocky pools. How still and beautiful they are, with their

* Phosphoridæ.

brilliant greens, and reds, and yellows, when, after lying closed like mere lumps of jelly, they open out into gorgeous flowers.

The sea-anemone really stands higher in life than the hydra and its companions, for the tube of its body is double, one end being doubled back within the other so as to make a small sac hanging within a large one, while a hole at the bottom of the little sac or stomach opens into the body-cavity below. The wall of the body between the two bags is divided into a number of narrow partitions (*s*, Fig. 25), upon the sides of which the eggs of the young anemones are formed, and out of which the tentacles spring as hollow tubes.

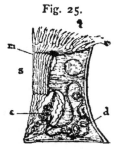

Fig. 25.

Section of a Sea-Anemone (*Gosse*), having special darts, *d*, which shoot out when it is attacked. *m*, Mouth. *s*, Stomach, showing the partitions on the sides of which the eggs are formed. *c*, Coil of lasso-threads in the stomach. *t*, Tentacles which are crowded with lasso-cells.

Yet it is more in the way of fleshy growth than in sagacity that the anemone has advanced, for in sensitiveness to light and power of movement he is far behind the floating jelly-fish. This is indeed to be expected, for in his quiet stay-at-home life he needs a strong muscular body, but not active senses, and so we find that while his lassos are powerful and many, his sight is only enough to lead him to move towards the light, and he shifts slowly along when he wishes to change his place, or floats with his disk upwards, without being able to choose his own path.

His fishing and feeding powers, on the contrary, are very great. Any one who has placed his finger among the tentacles of a sea-anemone will have felt how they cling to it, so that it is not always easy

Group of Anemones.

a a', **Painted Pufflet.***   *b*, Snake-locked anemone.†   *c*, Daisy **anemone**.‡
d, Cave anemone.§   *e*, Gem Pimplet.‖

to draw it away. The touch has in fact burst a number of lasso-cells, and the threads have pierced the flesh, though they are too fine to give pain. Mr. Gosse once cut off a piece of his own skin with a razor and put it to the tentacles of the dahlia anemone (*Tealia crassicornis*), and when he afterwards

---

* *Edwardsia calimorphia.*   † *Sagartia viduata.*
‡ *Sagartia bellis.*   § *Sagartia troglodytes.*   ‖ *Bunodes gemmacea.*

examined it under the microscope he found it full of lasso-threads, standing up like pins in the skin, and showing what wounds an anemone can inflict. Now, when we reflect what a large number of tentacles many anemones have (a full-grown daisy anemone has more than seven hundred), we see that they must possess an almost countless number of lasso-cells, and that small sea animals, such as shrimps, worms, mussels, sea-slugs, and young fish, must fall easy victims to the poisonous threads. Even if any creature is so well protected by its shell as to escape the darts, it is encircled by the numerous arms and thrust into the stomach, at the bottom of which it meets with another thick coil of lasso-threads (c, Fig. 25) which are soon fatal.*

In this way the sea-anemones obtain abundance of food, and they seem able to devour an almost unlimited amount. But they in their turn are evidently very open to attack, having such soft defenceless bodies; and in fact thousands of them must be devoured every day by sea-slugs and other animals, for they multiply so very rapidly that otherwise the whole shore would be covered with them. A sea-anemone can increase in three ways, either by splitting in half, or by throwing out buds, or, as is most common, by hatching the young from eggs within its body. It is most curious to see in an aquarium how quickly a crop of young sea-anemones springs up round the old ones. Mr. Holdsworth found that daisy anemones sometimes throw out as many as 146, 160, and even

* Mr. Charters White has told me of the case of a young fish struggling within the stomach of a sea-anemone and coming out uninjured; but such cases are rare, and may occur from some weakness or indolence in the particular anemone in question.

300, in one day. It is very difficult to see the young anemones born, because they are at first so small; but by careful watching they may be seen coming out through the mouth of their parent, sometimes in the shape of little hairy or ciliated swimming bodies, but more often as perfect tiny anemones which have lived inside their mother till their tentacles have grown. After they have been hatched among the partitions in the anemone's body, they generally travel into her hollow tentacles, and from there they are passed out through the mouth. Then after walking about a little while on the tips of their tiny arms they settle down and begin their life.

The first thing they learn to do is to expand to find food, and this they do by taking in water at their mouth or through their skin and so swelling out the whole body. But should an enemy come by they soon force the water out again, and become a small lump, very difficult to seize. It is most interesting to watch an anemone when it wishes to expand, gradually filling itself with water, and stretching its tender skin till each tentacle falls in its place as a graceful flexible tube, and then again in a moment, if you touch it, the water is squirted out, and every delicate part drawn in within its tough hide.

But if you touch a daisy anemone, or a cave-dwelling anemone, in this way you will find that it has another weapon of defence hidden in the body-tube itself. All the members of this family of anemones (*Sagartiadæ*) have minute slits scattered over the outside of their tube, and if you offend them these slits open and long white threads (*d*, Fig. 25) are shot out to strike you. These threads come from the coil of lassos within the body; they are not themselves lassos but long darts crowded with lasso-cells, and after they have punished the enemy that attacks them, they can be drawn in again to be used next time. By far the larger number of British anemones have these darts (called *Acontia*), so that we find even these sluggish stay-at-homes well able to fight the battle of life.

But mingled in among these soft lasso-throwers even on our English shores we find small examples of a still more wonderful race, whose history in the warm depths of the Mediterranean and amidst the stormy surge of the Pacific is like a fairy poem. Who has not heard of the groves of lovely red coral seen through the clear blue waters off the coasts of Corsica and Sardinia; or read of those islands which are built in the midst of the stormy Pacific by the delicate coral animal? There, in the midst of violent foaming breakers, strong circular stony reefs, crowned with delicate white sand and shaded by the cocoa-nut palm, enclose those peaceful lagoons where

> "Life in rare and beautiful forms
> Is sporting amid the bowers of stone,
> And is safe when the wrathful spirit of storms
> Has made the top of the waves his own."

And the coral-animal which builds alike the slender pink stem of the coral ornament, and whole islands of rock in the midst of the sea, is a lasso-thrower.

In the Mediterranean he is a delicate dainty being, beginning life as a little jelly-body thrown out of the mouth of a pure white polyp growing out of a red coral branch. This jelly-body soon settles down on the sea-bottom (*a*, Fig. 27), and spreading out its tentacles (*b*, Fig. 27) to feed, takes carbonate of lime from the water, and colouring it, we scarcely know how, begins to build with it red spikes or spicules into its jelly flesh; only into its mouth and stomach it lays no spicules but leaves them soft and white. Then after a while it begins to throw off buds, as we have seen the hydra do and some anemones, and each of these

Growth of Red Coral.*—*After Lacaze-Duthiers.*

a, A young coral settling down. *b*, The same putting out its tentacles. *c*, The same gradually forming new mouths. *d*, A coral branch with numerous mouths.

buds remains on the stem pure and white, while the jelly, full of red spicules, joins them all together (*c*, Fig. 27). And then as more and more buds are formed and the branches lengthen out the young coral becomes a coral-tree (*d*, Fig. 27), with all its buds or polypes spread out like dazzling pure white flowers, each with its eight rays expanded over the

* *Corallum rubrum.*

red jelly. Meanwhile in the middle of the stem the spicules become pressed together and form a solid red rod (*a*, Fig. 28), supporting the whole animal-tree; and this red rod, the scaffolding of the living lasso-throwing coral-animal, is all that remains after it is dead to be polished for us to wear.

All round the coasts of South Italy these beautiful coral-animals grow and feed. A warm sea and suffi-

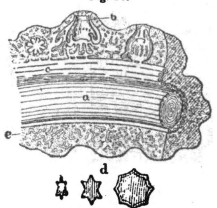

Fig. 28.

A section of a piece of Red Coral.—*Lacaze-Duthiers.*

a, Solid red rod in the centre. *b*, One of the white jelly mouths with its tentacles drawn in. *c*, Canals round the rod. *d*, Red spicules (magnified) which are buried in the flesh *e*. *e*, Soft flesh of the coral coloured by the red spicules and fed by the white mouths *b*.

cient water over their heads is all they ask, in order to flourish happily and send out plenty of young ones to keep up the colony; and though they have their enemies in the seaworms, and in the fish which nibble at their tender flesh, yet, by means of their spicules, they hold their own, while with their lassos they catch their prey.

A far hardier and more sturdy animal is the builder of the white coral, as he stands out in the midst of the wild Pacific, the stormy sea dashing against his home, while he has nothing but the power of life and growth to bring against it. Nevertheless, he not only lives, but builds strong stony barriers, which shut out the restless waves, and enclose calm, still salt lagoons, in whose depths more delicate corals can nestle and flourish.

Fig. 29.

Piece of White Coral.
(Madrepore.)

To understand how the white coral builds its skeleton we must look back to the sea-anemone, and to the partitions in the wall of its body (p. 67). The white coral is in fact a group of sea-anemones all growing together, and throwing out buds which remain on the stem, and each bud, as it takes the carbonate of lime out of the water, builds it up in solid layers between those partitions in its body. If you can find at the sea-side the little Devonshire cup-coral (Fig. 30), which is a single coral of this kind, you will be able to see clearly these solid partitions entirely enclosing the body. In this way the animal is fairly shut in, only the stomach with its mouth and tentacles remaining free; and as it buds and buds, feeding greedily with its lassos, and laying down lime particle by particle out of the restless sea, it builds a firm skeleton, sometimes branched (see Fig. 29), sometimes solid, as in the brain-coral, according to the way in which the buds are given off one from the other. And when the animal dies, instead of leaving only a smooth stem behind, it leaves each little cup of lime in the shape of its own body.

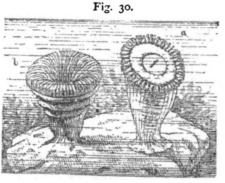

Fig. 30.

Devonshire Cup-Coral.*—*From Johnston.*
a, Living animal. *b*, Coral skeleton, showing the stony walls which the body lays down between the fleshy partitions.

How these corals have lived and grown for ages in the midst of the stormy Pacific, while the sinking bed of the sea carried down the dead coral as a solid wall, is a story which belongs to geology. Here we have only to picture the living animal, tiny and tender, yet strong in its two great powers—the power of catching and feeding on the creatures of the sea, and the power of building a solid skeleton with the grains of lime. In this way day by day, stretching out their tender arms and flinging their lassos by millions

* *Caryophyllium Smithii.*

and millions in the midst of the wild Pacific, the coral animals live and grow. In the midst of winds and storms they struggle on, the rough and strong builders without, in the open ocean, the more tender and delicate ones, with their bright coloured orange, crimson, scarlet, and purple tentacles, within the sheltered lagoons; they all make good use of the weapons with which life provides them, and flourish in countless numbers, enjoying the warmth of the tropical sea, and laying the foundation of solid rocks for ages to come.

This brings us to the end of our brief sketch of the lasso-throwers, of which the sea is so full. Though we have scarcely been able to glance even at the leading forms, we can understand how they are able to maintain their ground in the struggle for life. One and all, they sweep the waters with their tiny arms, and whether as animal-trees, jelly-fish, anemones, or corals, multiply in great numbers, and fill the sea with beautiful active life. If only as food for other animals, they have their great use in the world, for the huge whale is greatly dependent for his nourishment upon the shoals of jelly-fish which throng the Arctic ocean, and many shell-fish and other sea animals feed upon the anemones and delicate polypes on the sea-bottom. But beyond their use to others, is the great fact that they live and flourish themselves; like the rest of Life's children, they crowd into the world, and as we watch them during their brief career, we cannot but think that there is enjoyment in these fragile existences, as they open out so freely and eagerly in the depths of the quiet ocean; and that from them, too, rises the silent hymn of praise for the gift of life, even if it have its struggles and its dangers.

CHAPTER V.

HOW STAR-FISH WALK AND SEA-URCHINS GROW.

> "O, what an endlesse worke have I in hand,
> To count the sea's abundant progeny,
> Whose fruitfull seede farre passeth those in land,
> And also those which wonne in th' azure sky!
> For much more eath to tell the starres on hy,
> All be they endlesse seeme in estimation,
> Then to recount the sea's posterity,
> So fertile be the floods in generation,
> So huge their numbers, and so numberlesse their nation."
> — SPENSER.

ONCE upon a time, in a quiet sea-bay on the south shores of Great Britain, five curious little oval jelly bodies were swimming about by their jelly-lashes in the depths of the smooth water. They had one and all been hatched from eggs not long before, and their business and duty in life was to grow up into some form in which they could gain their living and protect themselves from harm.

As each one came from a parent of a different shape and character, it was natural that they should follow different roads, although they all worked much upon the same general plan; and though they were so small as to be scarcely visible, they soon began to put on each their own peculiar shape. No. 1 had not swum about for many hours before some lime-plates began to form in his body, arranging themselves in the shape of a cup (*a*, A, Fig 31), and below these other and smaller plates took up the form of a stalk (*b*). This went on for several days, while the jelly-body fed and swam about like any other living animal; but it proved after all to be only the cradle of the real creature, for after a time the jelly-body began to shrink up, and the whole sank to the bottom of the sea, and a strong lime-plate (*c*) was formed which fastened the lime-stalk to the rock, where the animal remained fixed, looking like a stony plant, and all that remained of the jelly was a thin film spread over the stem and cup. The jelly-animal had in fact become transformed into a *Crinoid* or Stone-Lily, about half-an-inch high, which soon put out jointed arms from its cup and fed in the water, and at this stage was a miniature copy of the well-known Medusa's Head,† which grows in the deep seas, and

Fig. 31.

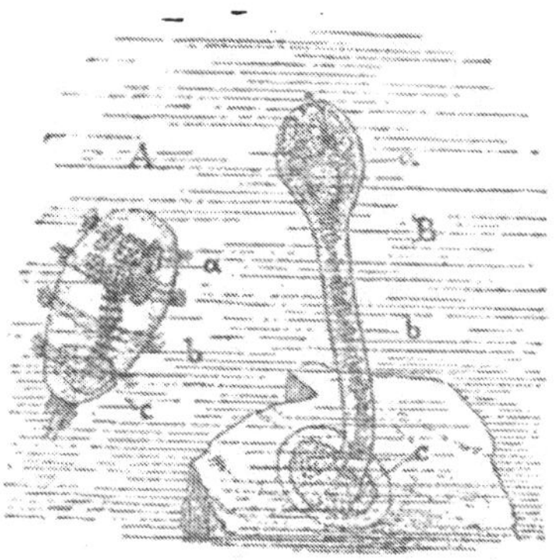

The infancy of a Feather-Star.*
Williamson.

A, The jelly-animal swimming by its lashes. *a*, The cup. *b*, The stem. *c*, The fixing plate of the young animal forming within.

B, The fixed animal from which the Feather-Star (Fig. 38) afterwards breaks off.

* These five figures, 31 to 35, are all much magnified.
† *Pentacrinus caput-medusæ.*

of those still larger Encrinites or Stone-Lilies, often more than five feet long, which we find fossil in the solid rocks of the earth, and which, though they look like the remains of stony plants were once true animals, feeding in the seas of ages long past by whirling the tiny sea-animals into the centre of the cup where their mouth lay turned upwards to the water.

No. 2 did not advance so fast, his jelly body had been from the beginning supported upon eight thin

Fig. 32.

The infancy of the Brittle Star-fish.—*Müller*.

A, The jelly-animal swimming and feeding while the Star-fish *b*, with its rays *c*, is forming inside it.

B, The young Brittle Star-fish which has swallowed the jelly and settled down upon the rock.

lime rods (*a a*, Fig. 32), causing him to swim along somewhat in the shape of a pyramid on legs, and he continued to float and feed in this shape for a considerable time. Meanwhile, just within his mouth some small cells appeared which gradually formed themselves into a round disc. By and by it was clear that a trellis-work of lime was forming over this disc (*b*, Fig. 32), and five tiny stony arms (*c*) began to grow out of it like the rays of a star. Still, however, the jelly-animal continued to feed through its jelly mouth like any other living being. Then after a time, during which there was built up within the disc a stomach, a mouth, and a set of tubes for taking in water, the disc with its sprouting arms all at once dropped off its rods and swallowed up the jelly-body, drawing it in till only a thin film was left over the stony star. Then, after swimming about for a little time, it settled down upon the rock and wriggled about, a tiny *Brittle Star-Fish* (B, Fig. 32).

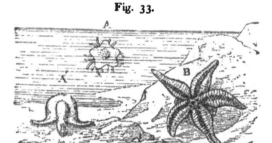

Fig. 33.

Infancy of the common Star-fish.—*Rymer Jones.*

A, Jelly-animal swimming about and the star-fish forming within it. A', The star-fish settling down. B, The same assuming its true shape.

No. 3 followed much the same course as No. 2, except that his jelly-body had no rods in it, but took a number of curious shapes and swam about briskly, while within was formed a young creature with a network of lime over his back (A, Fig. 33), and a number of small soft transparent tubes under his body. After a time the whole fell to the bottom of the sea, and this little creature also swallowed his jelly body, and becoming a tiny yellow rosette with five knobs sticking out of it, glided quickly away over the rocks, carried along by the little tubes under the rosette. It went on growing for two or three years, lengthening the five knobs into pointed rays, and became the common *Five-Fingered Star-fish*.

No. 4 took a different road from any of the three that had gone before him. He too had long thin rods in his body, all pointing one way, so that his body looked like a painter's easel, and at the top of the easel a number of fine plates of lime began to form in the shape of a tiny round box (*b*, Fig. 34 A) with prickles all over it; and by and by this box sucked up the jelly-body, leaving only a thin film over its shell, and sinking to the bottom a tiny *Sea-Urchin*, burrowed a hole for itself in the sand.

Lastly, No. 5 did not form anything solid within its jelly-body, but growing a stomach and feet, and other soft parts, stretched itself out into the shape of a sausage, put out some leaf-like tentacles round its mouth (B, Fig. 35), and laying down some spikes of lime in its skin, became a little worm-like creature with tiny tubes for feet, the young of the *Sea-Cucumber*,† and soon found

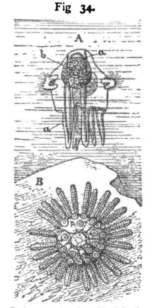

Fig 34.

Infancy of a Sea-Urchin.*
Müller.

A, The jelly-animal with its lime-rods *a a*, swimming about and feeding while the tiny sea-urchin *b*, is forming within it. B, The young sea-urchin.

* Echinus.   † Holothuriadæ.

some crack in the rock in which to hide its soft body.

These five animals—the stone-lily, the brittle-star, the common star-fish, the sea-urchin, and the sea-cucumber,—which grow up so curiously, each within an active feeding jelly being,* are the five types of the "Prickly-skinned" animals † which form the third division of the animal kingdom; and the history of their lives will give us a very fair idea of the implements and weapons used by this division, and of the peculiar walking apparatus which belongs almost exclusively to this branch of life's children.

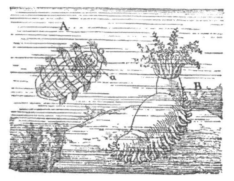

Fig. 35.

Infancy of a Sea-Cucumber.

A, A jelly-animal swimming and feeding; *a*, small sea-cucumber forming inside. B, The young sea-cucumber with the leaf-like tentacles round its mouth, walking on its tube feet.

Passing by, for a moment, forms 1 and 2, which we shall understand better presently, let us first visit the common star-fish after his arms are full grown, as we sometimes find him on the sand of the sea-shore thrown up by the waves. A strange and

---

* The jelly animal does not always swim about in the water while forming its future body. Some star-fishes and sea-urchins carry their young in a kind of pouch or tent till they have taken shape.

† *Echinodermata*, or hedgehog-skinned.

weird life he leads, clinging to the wet roofs and sides of caverns, or hiding under large stones, or wandering over the sand at low tide with the water rippling gently over his body: the sea must appear to him in a very different light from what it does to the coral-builders or jelly-fish, as they wave about their soft tentacles and bathe them in the element they love.

For the real interest of the star-fish is not in the sea above, but in the solid ground below. He cares for the water only that he may get oxygen out of it to breathe, for though he can swim when it is necessary, yet he is really a creeping animal, and loves to climb over the rocks, or poke about the sandy bottom with his mouth down to the ground, feeding on mussels and other shell-fish wherever he can find them.

No ghost could glide more smoothly or with less noise than he does as he wanders dreamily along, and when he comes to a wall of rock or a hollow in the sand he does not avoid them, but bends his body over the one or slides down the other, hugging the ground closely as he goes. And yet the machinery by which he moves is nowhere to be seen, nor will you be able to guess how it works, till you pick up the first living star-fish left upon the shore as the tide goes down, and put it into a glass pan or jar of salt water. Then you will be able to watch this curious movement through the glass. At first he will lie helplessly at the bottom, but very soon, although as you look down upon him you will not see any special part move, the whole body will begin to glide slowly along. Now lift the jar and look at the under part of the body. You will see hundreds of tiny transparent tubes moving in the groove under each of his five rays (A, Fig. 36, and *t t*, Fig. 37). The whole of the under part of his body will be waving like a field of corn, as each tube-foot in its turn is stretched out, bent forward, and fastened to the glass. Then after drawing the body a little on, it will loosen again and collapse into a mere knob, while another will lengthen out and take a hold. In this way, as tube after tube draws it forwards, the body of the star-fish will be

Fig. 36.

A, The common five-fingered Star-fish.* The dark round spot between the lower rays is the water-hole. B, The Brittle Star-fish.†

carried easily along the bottom or up the sides of the glass like a canopy resting upon the heads of more than two thousand bearers.

And now if you look in the centre of the under part of his body you will see a small opening with the skin puckered up round it. This is his mouth (*m*, Fig. 37), and if you place a small mussel or lim-

* *Uraster rubens.* † *Ophiocoma bellis.*

pet against the glass on his road you will see a curious sight. He will glide gently over it as though it were a mere stone, till his mouth is just above it, then the middle of the body will rise a little, and the feet all round the mouth fixing themselves firmly to the mussel will draw it into the opening, where it will remain till all its soft body is sucked out, and then the empty shell will return.

If, however, the shell-fish is too large to go into the mouth, the star-fish will apply its lips to it and often push its stomach-bag (S, Fig. 37) out at the opening and half cover its victim, and after a time when it draws back, the soft animal will be gone and only the shell remain.

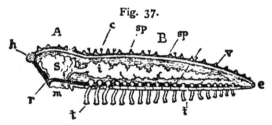

Fig. 37.

Section of the centre and one ray of a Star-fish.—*Rymer Jones*.

A, The central body. S, The stomach. *m*, The mouth. *h*, Perforated hole where water is taken in. *r*, Ring round the centre through which the water passes to the feet. B, The ray. *sp*, Spines set in the leathery coat. *c*, The snapping claws. *e*, Eye at end of the ray. *t t*, Tube feet. *v v*, Vesicles or waterbags supplying the tube feet with water. *i*, Liver.

The star-fish then is a kind of walking stomach, borne along by hundreds of tiny feet ceaselessly moving in each of its five rays, and it is the working of these feet which we must now explain. To picture to yourself the inside of a star-fish, imagine a round central dome-covered hall (A, Fig. 37), in the floor of which is a trap-door *m* (the mouth), and out of which open five stately arched corridors, one of which is shown in Fig. 37, which begin as lofty galleries and end in a point where a tiny window *e* is set. The roof and floor of the corridors are built of delicate white columns and arches of lime, joined by soft ligaments, while the walls are inlaid with star-like plates, and within the dome, and stretching right out into each corridor, lies the soft body of the animal (S, *i*, Fig. 37), with its digestive organs. The delicate telegraph of nerves, and the water-canal, starting from the central hall, pass like the wires and pipes of our houses under the floor of each corridor, while the numberless little water-bags which move the regiment of feet pierce the floor, and lie in the corridor itself.

And now, how does this apparatus work? Remembering as we do that the anemone spreads out its tentacles by filling them with water, we shall expect that something of this kind also happens here, only that we require besides to explain how the feet cling so firmly to the ground, for in some cases they will even break off from the body sooner than release their hold.

If you look carefully at the back of a starfish you will find a little round spot (see A, Fig. 36, and *h*, Fig. 37) lying at one side in the angle between two of the rays. This spot is a little plate of lime pierced with fine holes just like the rose of a watering-pot, and through it sea-water carefully filtered passes down a tube into a hollow ring (*r*) round the animal's mouth, and this ring opens again into canals which pass along under each of the rays. Here then we have a regular water-supply taken in at the porous plate and carried along all the five rays. But we want next a separate cistern for each tube-foot, for we have seen that they move separately, and so cannot all be filled with water at the same time. These separate cisterns we find in a number of elastic bags or *vesicles* (*v v*, Fig. 37) placed along the water-canals, and opening into them on the one hand, and on the other into the tube-feet. Now when water is taken in at the grating *h* above* the canals are filled, and they fill the vesicles, and each vesicle is able to contract and force its water down into its own foot-tube, thus stretching it out. Then the foot-tube while stretched at full length can, by drawing in its walls a little, force some water back, and so draw up the centre of the round cushion at the end of its tube, making a sucker just as a schoolboy does with wet leather on a pavement; then the foot holds fast. Lastly, by drawing up the muscles which run *down* the tube, the body is drawn on, the sucker released, and the foot pulled back to begin again.

This is how the star-fish walks, and when we remember how many hundreds of feet he has, how firmly each one holds, and how slightly it moves, we cease to wonder that he glides so smoothly and clings so firmly to the rock. He is a greedy creature, whose whole care is his stomach, and he will eat any animal food he can find, from small crabs, shell-fish, and other sea-creatures, to mere garbage and decaying matter, so that he is very useful as a scavenger of the sea. He in his turn is eaten by the cod, the haddock, and other fish, but he is better protected from smaller

* This grating is called the *Madreporiform tubercle*.

enemies than would appear at first sight. His thick skin contains a network of hard scales which will turn the edge of a knife if you try to cut it, while pointed spines (*sp*) stand in ridges on his back, and on the sides of the rays, thus protecting the tube-feet. But the most curious weapons he possesses are a number of minute claws, like birds' beaks mounted on stalks (*c*, Fig. 37), which stand round his spines, and twist and snap continually as long as he is alive. The only use that has yet been found for these curious weapons is to clear the skin of the star-fish from the seaweeds and small animals which would certainly fix themselves on such a sluggish animal if they were not picked off. We shall see presently in the sea-urchin that they are sometimes very active in this work.

And now as the star-fish plods on his way along the sea-bottom, thinking only of the creatures over which he can spread his capacious mouth, what has he to tell him of coming danger? How shall he be warned if the shadow of an enemy is passing over him, or if he is venturing too rashly into the broad sunlight where his bright colours might attract dangerous attention? If you notice any star-fish when it is alarmed or finds itself in strange quarters, you will see it curl up the tips of its rays, and there under the point of each ray (*e*, Fig. 37) may be seen with a magnifying glass a thick red spot seated on the extremity of a nerve, and having in it as many as from 100 to 200 crystal lenses surrounded by red cells.[*]
Here then we have a far better eye than that which we found in the jelly-fish, and it is no wonder that

[*] Haeckel, 1860.

the star-fish is so quick in finding food, or enrages the fishermen by discovering the bait which they put for other animals, for it turns out that this heavy, stupid-looking animal is much more wide-awake than he appears. In many cases a soft lid or feeler hangs over the eye-spot, giving to it a curiously intelligent look, and Professor Forbes relates how once when a beautifully delicate star-fish called the Lingthorn fell to pieces as he tried to lift it out of the water, this lid at the end of one of the arms "opened and closed with something exceedingly like a wink of derision."

Our first walking animal then is by no means a poor or feeble creature ; he has chain-armour woven into his leathery skin, with sharp spikes and snapping claws to protect him, a good digestion and a capacious mouth to feed his greedy stomach, a good array of nerves, quick feeling and eyesight, together with a wonderful apparatus for moving over the ground ; and when we add to this that if he loses any of his rays he can close over the wound and grow a new limb, we see that his powers of living satisfactorily are very great.

We must not suppose, however, that the curious walking apparatus of the star-fish is perfect in all his relations, or that they all walk by means of suckers, any more than all sponge-animals can build a toilet sponge, or all slime-animals make fine chambered shells. The rosy feather-star for example (Fig. 38), as it sits clasping the rock or a bunch of sea-weed, with the fine strong tendrils which grow out of its back, waving its arms like a group of brilliant red plumes spotted with bright yellow, has no need to use any feet-tubes, though it is a star-fish, and those which it has, probably serve merely as a help in breathing.

You will at first be puzzled to think how **this** feather-like fixed animal can be a star-fish at all, but if you examine it carefully, you will find that it is indeed one, only turned upside down. Its back, which is held down to the rock by some claws (*c*) which grow upon it, forms a cup in which lie the soft parts of its

Fig. 38.

The life of the Feather-Star.

A, Young of the Feather-Star before it has separated from its **stem**. B, Full-grown Feather-Star.* *r*, Rays. *c*, Claws. *m*, Mouth.

body, with a mouth (*m*), in the middle, of course turned upwards, and surrounded by tentacles. Its five arms have divided each into two, making ten stony jointed rays (*r*), and on these a number of finer jointed filaments give the appearance of feathers. Within a groove of each arm lie the feet-tubes, but they have no suckers, for the feather-star rarely walks, and then only wriggles in a clumsy manner, something

* *Antedon* (*Comatula*) *rosacea*.

like a brittle-star. It usually remains anchored, feeding on the minute beings in the water, which it drives into its mouth by hundreds of cilia or jelly-lashes which line the grooves of the arms.

It does not care to move at any time, though it can swim gracefully through the water when disturbed from its hold. But in its infancy it was not even free to do this, for the lovely feather-star is nothing more than the cup of the little stone-lily (A, Fig. 38 and Fig. 31), which has broken from its stem and grown up into a free animal. In the early spring you may find it in its infant state in the quiet bays of our west coast or of Ireland, like a white or yellow stony flower, growing on fronds of seaweed, or on small stony corals. Its stem of jointed plates is covered with a film of living matter, and its cup has the stony threads hanging down from it, which afterwards serve as claws to hold it to the rock. In the autumn you will find it so no more. The cup (*a*, Fig. 31, p. 78), floating off its stem (*b*) will have emancipated itself from the race of fixed stone-lilies, and joined the free star-fish, thus forming a curious link between these two groups of animals. It still, however, keeps much of its old habits, and while it can swim gracefully from place to place, loves better to cling to the nearest rock or weed, feeding upside down as compared to its new companions, and waving its deep red plumes, a harmless thing of beauty.

Not so, the brittle-star (B, Fig. 36), which, as we saw in Fig. 32, was a free being from the first, and is as voracious as the common star-fish, and much more active. In some ways, however, it is like the feather-star, for it has strong jointed suckerless arms and feet, which it never uses for walking, although it fills them through a porous plate like the star-fish. Its soft body too is all contained within the round cup in the centre, and its arms do not open out of it as in the star-fish, but are joined on; and this may partly explain why it so often flings its arms into a hundred pieces when frightened; for it can afford to part with them, and can soon form them again.

As tools and weapons, however, they are most useful, and the reason why the brittle-star does not use its tubes as feet is that its arms are quite sufficient to carry it along. Made of a number of small plates joined together by elastic muscles, and fringed with hooks and spines, these stony rays serve both as walking and feeding apparatus. The animal twists them to and fro in all manner of contortions, and in this way is carried over the rocks at a surprising pace, while it can bury itself in the sand and mud with the greatest ease, or wriggle into the smallest crevices if it fears to be attacked.

If the star-fish is remarkable for its smooth gliding motion, the brittle-star is the prince of wrigglers, and must escape many dangers by its bewildering activity. Indeed, we may almost fancy that its enemies may be as startled at its wild contortions as the fishermen were who dredged the brittle-stars up for Professor Edward Forbes, and begged to be allowed to throw them back, saying, "the things weren't altogether right!"

On the other hand, there is little doubt that they use their arms to carry food to the mouth, and one of this family called the "Basket-Fish,"* has its rays so branched and curled that they interlace, forming

---

* Shown in the left-hand corner of the Frontispiece.

a stony network in which crabs and small fishes are entangled and so caught for food.

Here we have then three types of prickly-skinned animals all bearing rays, and all having the same peculiar water-tubes, yet each of them has his own different life,—the feather-star, scarcely yet caring for his freedom, feeding almost in the same way as the polyps do among the lasso-throwers; the brittle-star with his active restless arms wriggling into cracks and seizing young crabs and shell-fish in his grasp; and the gliding star-fish with its thousands of tube-feet, creeping over its victims and carrying havoc wherever it goes.

But we have by no means yet exhausted the quaint designs of this ray-like structure; on the contrary, we come now to the most fantastic and whimsical creatures, not only among the tube-footed animals, but perhaps among all the inhabitants of the sea.

Is it because the sea-urchins know themselves to be as grotesque as the goblins of fairy tales, and as uncanny as rolled-up hedgehogs seen in the dim moonlight, that they hide themselves so persistently in the cracks of rocky pools, or bore holes in the limestone in which to hide their prickly bodies, or wrap themselves up in seaweed packed deftly between their spines? Or is it not more likely that they know too well the brittleness of their formidable looking spines, and either keep out of the way of the rolling waves and currents or protect themselves from their violence by a padding of soft seaweed?

Be this as it may, they are not always easy to

94 *LIFE AND HER CHILDREN.*

find alive, unless by those who know their haunts under large stones on the sand, or who fish for them in deep water; yet they are plentiful on all our coasts, and most people have picked up fragments of their shells upon the beach. When they are found, however, and placed in salt water, they well repay the trouble of a search, if only because they are so different from anything we have seen before.

Imagine a hedgehog rolled up tightly into a ball and beginning to walk along, not on his feet, but on the tips of his spines as if on stilts, and putting out here and there long fine tubes like threads of gutta-percha to anchor himself on his road, and you will have a fair picture of a walking *echinus* or sea-urchin, as he moves slowly along an aquarium or over the rocks on the sea-shore. There is something singularly whimsical in the movement of this prickly ball as it gravely lifts some of its sucker-feet to plant others, guiding itself the while by its movable spines. Each spine looks so knowing, turning itself round by its ball-and-socket joint, apparently making its own little excursions without regard to what the other spines are doing; and in large specimens, where the little claws can be seen round the spines

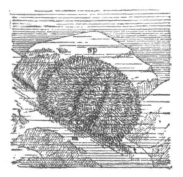

Fig. 39.

A Sea-Urchin * walking on a rock.
m, Mouth.   *t*, Walking tubes.
sp, Spines.

* *Echinus sphæra.*

twisting and snapping incessantly, the effect is more comical than can be expressed in a description.

But our sea-urchin is something more than amusing, he is a most wonderful example of how animals can be built upon the same plan, and yet so altered to suit their life that we should scarcely recognise them as relations. Looking at a sea-urchin, who would believe that it has anything in common with the starfish? Yet if you examine it without its spines, a rough description will soon explain how alike they are.

Suppose you were to take a dead star-fish and bend its rays backwards till they meet round the disc of the back; sew the tips there, and then sew the five rays together up the sides so as to form a ball flattened in the middle, you would then have the mouth of the animal (*m*, Figs 37 and 40) underneath the ball, and the five rows of feet (A, Fig. 36, *fh*, Fig. 40) running up it, while the edge of each ray where there are no feet would touch the edge of the next ray, making two rows of footless strips between each group of suckers.

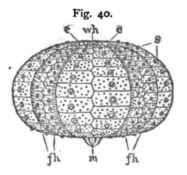

Fig. 40.

A Sea-Urchin after its spines have been rubbed off.

m, Mouth. *fh*, Foot-holes through which the walking tubes pass. *wh*, The water hole. *e*, Eyes. *s*, Sockets of larger spines.

If you could now blow out this ball so that the mouth and back were some distance apart and the whole was round, this would roughly represent our sea-urchin without its spines.

We see, then, that the tiny urchin which came out of the jelly-animal No. 4 grew up strictly according to the true ray-like (or *radiate*) plan, and yet what a change he has made and what a snug home he has formed round his body !

By laying down plates of lime within his soft flesh he has built a strong box, in which all his soft parts are enclosed, and at the same time has managed to keep a complete communication with the outer world. His sucker-feet, which act exactly like those of the starfish, lie safely within the box till he needs them, and then each one is put out at a tiny hole like a pinprick in one of the rows (*fh*). The porous plate (*wh*) supplying them with water is at the top of the shell in the back, where it would be in the star-fish, and in the other plates near it are the openings out of which it passes its eggs. But where are its eyes? Consider for a moment where they ought to be upon the star-fish plan. At the tip of each ray, and therefore, when the rays are turned up so that the tips meet round the back, they will be at the top of the shell, where you will find five small holes containing eyes (*e*), not so perfect as those of the star-fish but sufficient to see light.

Could a stronger or safer fortress have been designed even by the most ingenious engineer? No single soft spot is left bare to attack except the skin round the mouth, and this is always turned to the ground and defended by the spines projecting on all sides. The mouth itself is a most complicated piece of mechanism, with five strong teeth set in powerful jaws, which lie inside the shell.

And now how is this box to grow? The tiny

sea-urchin left nestling in the seaweed has to grow up to a large animal, sometimes as big as a pomegranate, and yet its body is tightly shut in within lime walls. Look again at the shell after it is stripped of its spines (Fig. 40), and you will see that it is made of more than a hundred separate plates. While the animal is living these plates are covered within and without by a slimy film, and this film passes also *between* each plate. Now as the animal grows it takes fresh lime from the sea-water into this film, and places it, atom by atom, evenly on the edges of the plates, and so the shell grows with the body without disturbing any part; and if this does not give sufficient room it can also add some plates to the top of the shell at the end of each ray.

So the sea-urchin lives and grows, wandering over the seaweed beds and grazing with his powerful jaws as a sheep grazes in a meadow. Though the shells of animals are sometimes found in his stomach they are not his proper food, for he is a vegetarian and might probably almost be said to chew the cud in his powerful jaws, which Aristotle called by the curious name of "lantern" from their peculiar shape.

He has many powerful enemies, and his shell is often found in the stomachs of large fish and other sea-animals; so that besides his strong box he has great need of his spines for protection, and he can give very sharp pricks with them from out of his hiding-places when he is interfered with. His spines, however, serve many other purposes. They guide him when he walks, they help him to burrow in the sand, they have even been seen passing seaweed and other objects over his body, and they help the little snapping claws to clear away any refuse which may gather on the shell. Lastly, the sea-urchin, which, like the star-fish, often protects its young ones in their soft infancy, will sometimes gather the spines together at the top of its house, and so form a tent for the tender young urchins till they are fit to go alone.

The snapping claws, which we found before in the star-fish, exist in numbers on the shell of the Echinus, and are very puzzling; they are so very active and yet seem to do so little work. They have often, however, been seen passing away the little pellets of refuse food which come out of a hole in the top of the shell. These pellets are handed down from claw to claw till they can be dropped into the water and so got rid of. In the same way small worms and seeds of plants and other living things are cleared off the bristling shell by these busy little snapping beaks. The spines by their constant movement help, as we have seen, in this cleaning process, and have probably many uses not known to us.

Who will say when he examines the structure and studies the habits of the Echinus, that this child of life is not a quaint, clever, wonderful, and skilful piece of mechanism, as it lives and breeds by thousands in the depths of the sea? Any handful of seaweed out of a pool at low tide will contain some, so small as hardly to be noticed; while from the rocky depths of the Mediterranean the fishermen bring up large ones in order to sell their bunches of eggs for food. Yet, as they stand in the Italian markets feebly moving their spines round and round in search of some of the old familiar objects in their sea-home, how few people stop to examine the curious box or to think of the history of its dying architect!

And now, what has been happening all this time to the small worm-like creature No. 5, which we left hidden in the rocks? You will have to search well in the crevice of some dripping cavern only lately deserted by the tide, and there you may perchance find him bathing himself in a rocky pool, a large soft satiny sausage, purple, white, or brown, with five delicate stripes down his body (see Frontispiece), and a wreath of beautiful purple tentacles like fine seaweed waving round his mouth. What connection can this worm-like creature have with the rayed animals?

Wait awhile and look more closely. Sluggish though he is, the Sea-cucumber does care to move sometimes, if only to fill his body with sand, and so get the particles of living matter which form his sole food. As he begins to glide along, see, from the five stripes running along his body there appear a number of tiny tubes with suckers (see Frontispiece),* by which he draws himself along. Here then are again our five rays of tube-feet, but this time not forming a star, or gathered into a ball, but stretched out along a soft fleshy tube.

It would seem almost as if here life had neglected to arm the poor soft Sea-cucumber, or, tired of inventing prickly defences, had fallen back again upon a soft jelly-animal. But the creature is not so helpless as he appears, for in his thick transparent skin are strong muscles by which he can draw his

---

* In some of the sea-cucumbers the rows of feet are all drawn together on the under side of the body, and this is the case in the form which the artist has represented on p. 82.

body in and out much as a worm does; and some species have sharp hooks buried in their flesh which both help them in moving and in wounding those who attack them. But his great safeguard is his power of contraction. Try some day to find a sea-cucumber in a crevice on the sea-shore, and then set to work to get him out. You will feel him slip through your fingers like an eel, as he squeezes the water out of his body and forces himself into a narrow crack from which he cannot be dislodged without breaking the rock. There is a safety in pliability which is sometimes surer than a stout resistance, and where the prickly sea-urchin might fall a victim, the sea-cucumber effaces himself and escapes.

A curious mixture he is of the savage and the cultivated animal. Though he gorges himself with sand, which seems after all but a coarse way of getting a living, yet his body is more delicately formed than that of any other prickly-skinned animal, and this makes it all the more strange that he should have the power of throwing out nearly the whole of his inside, and yet living and growing it again. Sir John Dalyell found that a sea-cucumber which had lost its tentacles, its throat, its network of blood-vessels, its intestines, and its egg-sac, and had literally nothing left but an empty tube, lived, and in three or four months had regrown all the inside of its body. An animal which can exist like this, and is scarcely ever found with all its parts complete, because it has parted with some of them, and yet is healthy and strong, need surely not envy the brittle-star its stony case and wriggling arms, nor the sea-urchin its strong box.

And now we have followed our five little jelly-bodies out into life, and have found that they have as much a real history as you or I have, with real struggles and difficulties which they can only overcome by using all their powers. The varieties of these five forms are far too many for us even to glance at them. There are the fixed stone-lilies of the deep sea, which do not become free like the feather-star. There are brittle-stars, from a tiny star with a disc as small as a pin's head, and arms like fine threads, to others measuring a foot and a half across. There are star-fish large and small, some like stars, others like five-sided plates, others with the rays turned back like a folded dinner-napkin. There are sea-urchins round, egg-shaped, wheel-shaped, and flattened, and from the size of a pea to that of a child's head; while there are others from warm seas, with three-edged spines as thick as a little finger, and twice as long. A visit to any good museum* will show these varying forms, and though the sea-cucumbers will not be so well represented, because they are soft animals, yet you will find the Trepangs of the Chinese, with their black leathery coats, and others which are covered with plates of lime. The beautiful Synapta, which lives in our English Channel, with its lovely rose-coloured tube, and its anchor-bearing shields, you will not so easily find, for it is so brittle that it is very difficult to preserve. This lovely creature, often a foot and a half long, shelters itself by a tube of sand built in rings by its tentacles, and passed down over its body by the microscopic anchors buried in its soft flesh, and by these anchors it also

* The British Museum has a very fine collection.

draws itself in and out, showing a new expedient used by an animal in which the tube-feet are wanting.

These and many other wonderful adaptations are open to all to study, but we must not linger over them here. One marked step we have made in this division—we have advanced from mere floating or fixed animals to creatures able to wander freely over the floor of the ocean. The children of life have now got their feet upon the ground, but not yet their heads above water. In fact they have as yet no heads to put anywhere. Eyes, ears, mouths, and feet we have met with, but no heads, nor have any of these animals been able to live out of their watery home.

But soon a new prospect opens before us, and in the mollusca or soft-bodied animals, and the worms, we shall begin to enter upon earth-life. Not suddenly, however, for all new powers are of slow growth, and through many chapters yet we shall find the largest number of each group clinging to their old ocean home, and only here and there air-breathing and head-crowned forms mingling in the throng.

CHAPTER VI.

THE MANTLE-COVERED ANIMALS, AND HOW THEY LIVE WITH HEADS AND WITHOUT THEM.*

See what a lovely shell,
Small and pure as a pearl,
Lying close to my foot,
Frail, but a work divine,
Made so fairly well,
With delicate spire and whorl,
How exquisitely minute,
A miracle of design.

The tiny cell is forlorn,
Void of the little living will
That made it stir on the shore;
Did he stand at the diamond door
Of his house in a rainbow frill?
Did he push when he was uncurled,
A golden foot or a fairy horn,
Thro' his dim water-world?

TENNYSON.

OF all our many playthings when we were children, were there any we loved better or cherished longer than the shells which we brought home from the seaside, and each of which we knew, not perhaps by name, but as a shepherd knows his sheep, so that no single one could be missing without our detecting it?

They may have been only common shells, such as the small pink-tinted scallops, variegated top-shells, small cowries, or spiral turrets, with here and there a delicate razor-shell, treasured espe-

---
* The sea-mats (*Polyzoa*), sea-squirts (*Ascidians*), and lampshells

cially because so easily broken. Yet we felt instinctively that they were more beautiful than any artificial toys, and though probably we scarcely thought of the animal which formed them, yet the delicate marking and tints of colour which each had left upon his house, pleased our eye more than gaudy pictures or painted playthings.

And even amongst older people is there any place in the world where shells are not admired? The savage strings them into necklaces, and so does the refined lady of fashion; while there is probably not a house, even the poorest in England, where they do not figure as ornaments, from the giant conchs and cowries of the South Seas, brought home by some sailor son, to the little boxes made of our common coast shells.

Now each one of these millions of shells preserved in all parts of the world, as well as of the countless multitudes which lie crushed and broken on the sea-shore and at the bottom of the sea, has once been the home of a living animal, which was born wrapped in a transparent mantle endowed with the wonderful power of extracting lime from the sea-water which it has taken into its body, tinting it with beautiful colours, and building it up into a solid house.

This wonder-working mantle which life has given to these soft-bodied mollusca (*mollis*, soft) may easily be seen in any common shell-inhabiting animal, such as the oyster or the periwinkle. When an oyster is

(*Brachiopoda*), are purposely omitted in this chapter, because although familiar objects, yet their structure is too difficult and their true position too uncertain for them to be dealt with in a book of this kind.

opened you may see two transparent flaps, with thickened edges, one lying above, and the other below the oyster in its shell (*m*, Fig. 41, p. 108), and these two flaps are the two halves of the mantle, which, when they touch, enclose the animal between them. In the periwinkle the mantle is equally visible, but this time it is all in one piece, and forms a complete transparent tube, out of which the animal pokes its head and its crumpled foot bearing the horny lid, or *operculum*, which closes the shell.

When the periwinkle was very young he was not larger than the head of a small pin, and his shell was like a minute transparent bead. But as his body grew larger it was necessary for his home to be both larger and stronger. Then he stretched out his mantle till it reached over the edge of the tiny shell, and gave out from it a thin film, in which were grains of lime which had been passed through his body into the mantle. This film, clinging to the inside of the shell and stretching over its edge, formed a fresh internal layer, and a new rim to the mouth. The rim, however, was not white, but coloured by little cells of dark paint or pigment, secreted in the border of the mantle. The shell was now a little larger and a little thicker, and the mantle was drawn in till a still more roomy house was needed, and then the same thing took place again; and so the building went on till the shell was completed, the lines round and round it marking the rims which had each in their turn formed its mouth.

In this way the mantle, not only of the periwinkle but of all the mollusca builds up the shell for the animal to live in. In the oyster each half of the mantle lays down its own separate valve, and this is the case with all those mollusca which have no heads; they all grow *bivalve*, or two-valved, shells, while those which have heads, such as periwinkles, snails, and whelks, have their mantle all in one piece, and consequently grow single or *univalve* shells.

Nor is this all, for the shape, colour, and peculiarities of all the different shells come from peculiarities of the mantle. If this is crumpled at the edge or drawn out in horn-like folds, then the shell will have a crumpled form like the scallop, or horns like the murex, while the sunlight falling upon the mantle seems to help it in forming the bright pigment with which it paints its home, so that shallow-water shells and those of the tropics are more brightly coloured than those from the deep sea or from dull climates. Again in the inside of the shell, if the mantle leaves a smooth layer this will be white, but when the film is crumpled in very fine folds, these reflect the light in such a way as to give the beautiful colours known as mother-of-pearl; while, if the mantle be irritated at any point, it will form in the oyster or the mussel a little bead of lime afterwards to be increased into a pearl.

And now with this picture in your mind of the mantle at work, visit any good collection of shells, such as that at the British Museum, and look at the giant Strombs and Volutes of the Indian Ocean; the Pinnas from the Mediterranean, half a yard long, with their erect curled scales; the prickly Murex with its delicate pink-tinted lining; and the gorgeous purple Mussels. Notice the rainbow-coloured chambers of the Nautilus, the pearly lining of the Haliotis, and the lustrous transparent shell of the floating Carinaria, and then say whether the work done by the mantle of the soft-bodied animals does not surpass that of any sculptor or artist in the world!

Yet this is not the chief object of the shell, which is meant to shield the delicate creature within, and does it so successfully that though the soft bodies of the mollusca offer the most tempting morsels to birds and insects on the land, and to almost all the inhabitants of the sea, yet, protected by their shelly covering they spread into every nook and corner of the globe, giving birth to such multitudes of young, that, in spite of all the havoc which thins their ranks, they flourish in abundance. Even the

"Poor patient oyster where it sleeps
Within its pearly house,"

although it is the most helpless of all the headless mollusca, would overspread all the deep-sea banks round our coast if we would let it alone. The oyster fishers have only to visit their well-known haunts about half-a-mile or a mile from the shore, in Essex, Kent, Wales, and elsewhere, to rake them up by hundreds. If you could dive down there to the bottom of the sea you would find the oysters cemented firmly to the rocks and to each other by the under part of the largest valve, which is cup-like in the centre where the body lies, while the flatter valve is gaping open and a stream of water is gently flowing over the oyster within.

The shells stand naturally open because they have an elastic cushion (*c*) something like a thick piece of gutta-percha fixed within the hinge, which acts like a spring of a jack-in-the-box, and drives the covering valve up unless it is forcibly pulled down. This, however, can be done by a strong muscle (*ms*) which lies within the valves, and has one end fastened to the upper and the other to the lower valve, so that by contracting this muscle the oyster can pull its shell together with a snap when danger is near. Close round this muscle lies the body of the animal between the two flaps of the mantle (*m m'*). Lifting up the upper flap you will find, edging the body and growing to the mantle, a delicate transparent frill (*g*) of four striped bands, these are his gills or breathing apparatus.

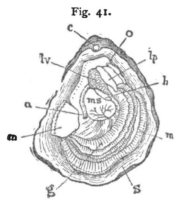

Fig. 41.

An Oyster * lying in the shell.

s, Shell; *m*, lower half of mantle; *m'*, a piece of the upper half; *g*, breathing gills; *h*, heart; *lv*, liver; *lp*, lips; *o*, opening of mouth; *a*, anus where refuse is thrown out; *ms*, muscle holding shells together; *c*, elastic cushion forcing them apart.

> " The fringes that circle its body,
> Which epicures think should be cleared,
> Are the animal's lungs—for 'tis odd, he
> Like a foreigner breathes through his beard."

The stripes are tubes which run up and down each fold, and through them flows the sluggish colourless blood of the oyster, so that as the gills lie bathed in water, the blood takes in oxygen through the delicate membrane, and flows back to the body purified and refreshed. The remainder of the oyster

* *Ostrea edulis.*

consists of its stomach, digestive tube, and dark coloured liver (*lv*), an ovary where the oyster eggs are formed, and a heart (*h*), with two chambers, which pumps the blood through the channels of the body, while fine nerves spread in all directions, not yet arranged in pairs along a cord as we shall find them afterwards in insects, but straggling to the various parts from two chief centres.

But where is the mouth? Placing the oyster with its deep shell downwards, and the rounded part towards you, you will find an opening (*o*) in the right hand corner near the hinge, and over it four thin lips (*lp*).* If you could watch the oyster alive, you would see that all the water passing over the gills flows towards this mouth, and the reason is made clear if you put a small piece of a gill in water under the microscope; for then you will see a whole forest of lashes waving over the surface of the gills like reeds in a stream, and striking strongly in one direction, namely, towards where the mouth would be. By means of the action of these lashes, or *cilia*, the oyster, as he lies gaping in the water, has a constant current flowing over him, which not only provides him with breath, but drives the helpless microscopic plants and animals past his thin lips, to be drawn in and swallowed.

But though the oyster has little trouble in obtaining his food, he has much in preserving himself from danger. When he first comes out of the egg, he remains for some time lying safely between the gills of

* In opening oysters at the shops, they turn them on the flat valve, and remove the round one, so that the mouth will then be seen on the left side.

his parent, but by and by he is cast out, to make room for others (for one oyster may lay as many as two million eggs in a year), and swims away by means of a number of lashes, which extend beyond his thin transparent shell. Woe betide him then if he comes near to a hungry fish, or crab, or sea-anemone, for millions of young oysters are swallowed by these animals; yet he is not quite without help, for at this time he has two little red eye-spots, and can see his danger, whereas he loses these after he is fixed to the rock. Still even then his nerves seem sensitive to light, for his valves are said to close at once when a boat passes over him in clear water, and his sense of touch is very acute all round his mantle; and as he builds his shell firm and strong, he can show fight against many intruders, and live sometimes for ten, twelve, or fifteen years.

But it is amid many perils, for the star-fish can apply his greedy mouth to the valves, and stifle him in his grasp, and annelids or sea-worms can work their way into his shell, while the whelk with his rasping tongue bores right through it, and feeds on his tender flesh; and, if he escapes all these, the boring sponges destroy hundreds of his race by riddling the shells with holes, and growing upon the graves of their victims. Even his own children often cause his death, by settling down upon his upper valve, so that when a bank becomes densely populated those underneath are stifled in a living grave.

From the oyster which lives on banks at many fathoms depth, we will pass on to the mussel anchored nearer to the shore on the mud-banks and groynes which are uncovered at low water. Here the waves

THE MANTLE-COVERED ANIMALS.

beat roughly, and to be safe it is necessary to withstand them. But the mussels (M, Fig. 42) do not, like oysters, cement themselves down for life. They have a different stratagem which enables them to get free if they wish. They have below their body a muscular flap, which goes by the name of a "foot," and is made up of layers of muscle crossing and

Fig. 42.

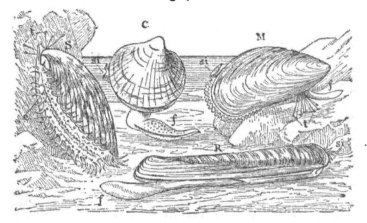

Group of headless Mollusca.
C, Cockle * springing. M, Mussel.† S, Scallop.‡ R, Razor-fish.§
f, Foot ; *t*, anchoring threads ; *si*, breathing siphons ; *e*, eyes of scallop.

recrossing each other. In this foot there is a deep groove, out of which they force a milky fluid which hardens into threads (*t*) and anchors them to the rock. Any one who has tried to wrench mussels from their bed, knows how strongly these threads hold ; and if you remove the mussels carefully and put them in an aquarium, you may see them anchor

* Cardium.   † Mytilus.   ‡ Pecten.   § Solen.

themselves. As soon as they grow a little accustomed to the place they will begin feeling about with their foot to find a spot, and then pressing the tip firmly against it, will draw it back after a time, leaving a thread behind. The huge fan-mussel or *pinna*, common off Plymouth, forms threads so silky that they have actually been woven into gloves. The mussel then has the power of spinning new threads and settling in new spots, but he is practically a stationary animal, providing himself with plenty of food by the rapid motion of his fringed gills, so that even young shrimps in spite of all their efforts are carried into the whirlpool. Then when the tide goes down, he closes his shell, shutting in enough water to last till the sea returns, and it is while he is left high and dry that the sea-birds often wrench him from the rocks and devour him.

In the scallop (S, Fig. 42) we get a step farther; for though he too forms a slight cable and anchors himself to the rock, yet he can in most cases withdraw it at will and dart through the water in long rapid leaps, so that a group of young scallops look as if they were performing a dance. Mr. Gosse, who watched this in an aquarium, saw the scallop draw as much water as it could hold within its mantle, and then, closing the edge, squirt it out at one corner so as to drive itself along in the opposite direction. The lima, which is nearly related to the scallop, and has a lovely orange fringe to its mantle, often builds a nest with its threads, working in pieces of coral, gravel, and shells, and fastens it to the seaweed, lining it with a smooth layer of slime, and taking refuge in it out of the way of crabs and fishes. But the scallop goes boldly out into the sea, and you will not wonder at its activity when you see its beautiful jewelled eyes (*e*) set all round the rim of its mantle like precious stones set in a ring. You may easily see these eyes peeping out at you through the half-opened shell in any fishmonger's shop, and a pretty sight it is.

The life of the cockle (C, Fig. 42) is very different. True he can leap to a great distance by bending his long foot (*f*) and straightening it with a jerk; but he uses it chiefly to burrow in the soft sand, and then he draws his body down till only the tip of his shell is uncovered, and there he takes in water and food. Some cockles have the two flaps of their mantles joined together and drawn out on the side opposite the foot into two short tubes (*si*, Fig. 42), down one of which the water enters, while it is thrown out at the other.

Lastly, the razor-fish, whose shells we find so often, but whose bodies we rarely see, scarcely ever come above ground at all, but burrow with their thick foot till only the two siphons (*si*) are uncovered, and throw up jets of water, by which the fishermen find them when they dig them up for bait.

We have bivalves then lying fixed in the deep water, anchored on the stormy shore, and buried in the sand, nay more, if we search at low tide we may often find the rocks riddled with holes, and, on breaking them open, see within a Pholas, an animal like the razor-fish, but much shorter and with a beautiful delicate shell. The Pholas has learnt to find a home in the solid rock, while the groynes of our shores and the bottoms of our ships are destroyed by another true bivalve, the Teredo, which is miscalled a "shipworm."

Then we can trace these headless mollusca from their ocean-home gradually up into the fresh water, some forms living in the brackish water at the river's mouth, others like the fresh-water mussel buried in the mud of rivers; and these do not spin threads, since they have no rude waves to meet, but put out two short siphons to the pure water above. All kinds of different forms with their habits we may study on the coasts and in the ponds and rivers; but we never find a bivalve either on the land, or sailing in the open ocean.

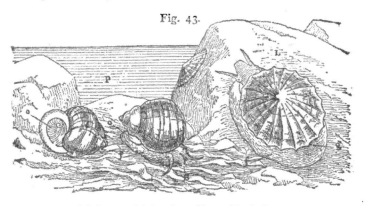

Fig. 43.

Molluscs with heads. Vegetable-feeders.

L, Limpet* walking, and attached. P, Periwinkle† walking, and closed. *f*, Foot; *o*, operculum; *s*, snout; *g*, place where gills lie under the shell.

These regions they are obliged to leave to the more highly-gifted mollusca with heads; and when we have examined the little periwinkle grazing on the seaweed among the rocks, we shall, I think, be able to imagine how it was possible for some of his

* *Patella vulgaris.*   † *Littorina littorea.*

THE MANTLE-COVERED ANIMALS. 115

distant relations to venture into new hunting grounds and become land animals.

Watch a periwinkle some day in his home among the rocks, and see him gently lift his shell, open his horny door (*o*, Fig. 43), and put out his head. He has two delicate tentacles to feel with, and just behind these on very short stalks are set two tiny but keen eyes, the nerves of which join the great nervous mass now for the first time chiefly centred in a head. The under part of his body is a flat crumpled disk or *foot*, as it is called, composed of muscles; and this when lengthened out first on one side and then on the other, draws him gently along, the under side being moistened from time to time by slime from a gland within. On account of this foot being under the body, the periwinkle and his companions are called stomach-footed (*Gasteropoda*). So he moves on, but at the slightest alarm he disappears as if by magic into his shell, drawing his horny door close be-

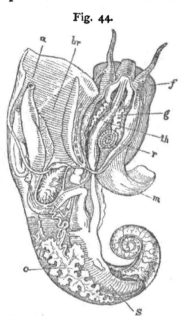

Fig. 44.

The inside of a Periwinkle.—*Bronn*.

f, Foot; *m*, muscle for drawing back into the shell; *g*, spittle glands; the glands for giving out slime are near the anus tube; *th*, throat leading to *s*, stomach; *r*, rasp of teeth rolled up; *br*, branchiæ or breathing gills, which, when the mantle is folded back in its place, lie over the throat; *a*, anus; *o*, ovary carrying eggs.

hind him, for the powerful muscles of his mantle (*m*, Fig. 44) enable him to shorten or lengthen his body at will. If undisturbed, however, he finds his way to a mass of seaweed, pushes out his snout (*s*, Fig. 43), and moves very slowly along, scraping fine shavings off the weed as he goes, so as to leave minute dents behind him.

This he does by means of a very curious instrument. If you could look into his mouth, which opens on the under side of his head, you would find it paved with sharp teeth, just as if a number of nails had been driven into it point upwards, and it is with these that he **rasps** the seaweed as he rubs his jaw along it.

But this rough file wears away rapidly with constant use, and to meet this difficulty he has a complete provision hidden within. The rasp within his mouth is only the end of 600 rows of teeth, three in a row, growing on a long gristly strap like pins stuck in a pincushion, and this strap, often two and a half inches long, closes its edges together at the back of the mouth so as to wrap over the rough points, and is then rolled up into a coil, and stowed away in a fold of the neck (*r*, Fig. 44). As the front teeth wear away this strap comes gradually forwards on the floor of the mouth, the new teeth grow up and are sharpened, ready for use. This curious strap is generally called the "tongue," though a "rasp" (*radula*) is a much more appropriate name.

And now as our periwinkle walks and feeds he must also breathe, and, strange as it may seem to us, no creature below the back-boned animals ever breathes through its mouth.* Look back to the earlier groups and you will see that the sponges, jelly-fish, and corals breathe through the skin, while the star-fish takes in water, not through his mouth, but through the perforated plate in his back; the oyster breathes by means of gills fringing his body, and we shall find by and by that insects breathe through holes in their sides. We must look then for the gills of the periwinkle, and we find them safely lodged in a fold of his mantle over his neck, just within the broad part of his shell (*br*, Fig. 44). There they are bathed in water drawn in by their waving lashes, and when the periwinkle is left high and dry by the tide he pulls-to his lid, shutting in a supply of water.

The same is true of the limpet, not that he has any door to close, but he clings so closely to the rock that water is shut in all round his gills, which fringe his body just above the foot. You would hardly imagine at first that a limpet has a head like a periwinkle, but when he is covered by the water and not afraid that the birds will peck at his tender foot and carry him off for food, you may see him lift his shell and put out his head with its horns, and make a track off to the nearest seaweed, where he grazes steadily. But when the tide goes down you will find him back again in exactly the old spot, where he has worn a little basin for himself to lie in, to which he fits so closely that sometimes his shell will

* Exception may be taken to this generalisation as regards the ascidians, but it must be remembered that, so far as the true nature of these has been determined, they appear to be degraded members of the vertebrate type.

have even grown a little deeper on one side than on the other to fit some dent in his nook.

These are the peaceful vegetable-feeders, and the margins of their shells (when unbroken) have always unnotched rims, but if you pick up a shell which has a notch (*n*, Fig. 45) in the margin as in a cowry

Fig. 45.

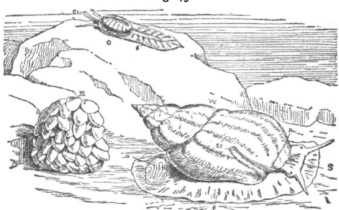

Flesh-feeding Molluscs.
W, Whelk. E, Whelk-eggs. C, Cowry.
o, Operculum ; *n*, notch in shell ; *si*, siphon ; *f*, foot ; *s*, snout.

or whelk shell, you may be almost certain that its owner fed on other animals, for flesh-feeders have their mantle folded right round their gills, and drawn out into a tube or siphon (*si*) through which the water is taken in, and the notch in the shell marks the place where the siphon protrudes.

Now these flesh-feeders have to work much harder for their living than the grazing limpet or periwinkle. Though they sometimes devour fish and other soft animals, yet their chief food is shellfish, and they have to reach them through their closed houses. The hungry whelk therefore has to bore a hole through a solid shell before he can take his meal, and for this he is provided with a boring instrument such as any engineer might envy. His snout, which can be stretched out like the trunk of an elephant, contains a toothed rasp like the periwinkle's but much more formidable; and this rasp is moved up and down by powerful muscles so as to act like a fine saw drilling a neat round hole even in the hardest shell, through which he can suck out the soft body it contains. It is curious that he does not always know when he will find food within, for he will sometimes drill a hole not only in an empty shell, but even in a shell-like stone.

While the periwinkle and his relations then are grazing on the seaweed, the whelks and cowries, and their tribe, are finding means to attack the oysters and cockles, limpets and periwinkles, and so to establish a successful hunting-ground where there would be no room for more vegetable-feeders; and you can scarcely pick up a handful of shells without finding some pierced with the holes made by these marauders. They people the shores of the ocean all over the world, some carrying their eggs till they are hatched, some glueing them down in safe nooks, others, such as the whelk, laying them in a bunch of horny bags (E, Fig. 45), in each of which the young whelk may be seen moving, if you can pick them up fresh from the sea. And when the little ones are born, they are able to swim about, as the young oyster was, and while myriads are borne away on the sea and devoured by other animals, the remainder settle down and feed on the sea-bottom.

This is the history of the sea-forms, and we have now to glance at those on the land. First, we must notice, in passing, the water-snails in the ponds and rivers, feeding on decayed leaves and travelling often from place to place, floating shell downwards on the surface of the water, or hanging from the water-plants by slimy threads. Some of these have and some have not the horny door, while some breathe by gills, and others are air-breathers. Then we have not much difficulty in recognising the land-snails as being very like the periwinkle, only breathing by air instead of by water. The way this is done is very simple. If you watch a snail when its head is out of its shell, you will see a little slit opening and shutting steadily in the top of the neck, and through this hole air is passing into a closed chamber made by a fold of the mantle. The walls of this chamber are covered with a network of blood-vessels, through which the blood flows, taking oxygen this time from the air instead of from water. By this simple arrangement the snail, no longer confined to the sea and rivers, is able to spread over the fields, and woods, and gardens, feeding on the delicate juicy leaves of plants, on mosses, and fungi, and all the rich vegetation of the country. But it has many dangers, for birds and hedgehogs, and even insects, prey upon it greedily. Therefore it feeds chiefly in the dusk of the evening; while it has sharp eyes (*e*, Fig. 46) set upon long stalks, which can see on all sides when it is out of its shell.

Now in order to retire safely into its shell, it must be able to draw in these eyes, and also the two tentacles or feelers below, and here we find a beautiful machinery. If you watch a snail drawing in its horns you will see that the eye disappears down the tube, just as the tip of a glove-finger does, when you draw it down from inside the glove. These horns are in fact hollow tubes, and a special muscle pulls them in from the top downwards, and when the eye is wanted again, it is only necessary for the muscles round the tube to contract, and so to squeeze the tip gradually out.

Most of the land-snails have lost the horny door, not having any need for it; but in winter, when they sleep without food in the cracks of old walls, under the bark of trees, and in other sheltered spots, they pour out a layer of slime, which hardens and shuts them into their shell till spring returns.

Slugs (C, Fig. 46), on the other hand, bury themselves in the ground for winter safety. At first sight you might imagine that a slug had no shell at all, but if you examine carefully you will find a small shell (*s*) *under* its black skin, just behind the neck, and the small breathing hole (*b*) at the side will show you that this shell covers the breathing organs. This is in fact the only part of a slug's body which is covered by the mantle, and if you alarm him you will see him draw his head in under it, as though he expected it to shield him from danger. No doubt the absence of a large shell enables the slug to creep into many places where a snail cannot go, and the havoc worked by these creatures in our gardens shows how rapidly and successfully they feed. The great gray slug* has a supply of 28,000 teeth, so

* *Limax maximus.*

that he can use them without scruple ; and if it were not for the birds which devour both slugs and snails at their work, and some insects which destroy their eggs, the whole land would be eaten up by them ; for they hide their eggs so cunningly in the roots of plants, in crevices, and well-sheltered nooks, that they multiply by millions.

Fig. 46.

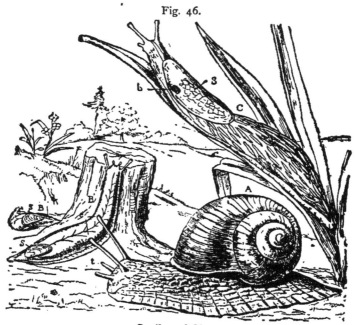

Snails and Slugs.

A, Garden snail.* B B, Testacella ;† one disappearing into the ground, and only the tail showing. C, The Great Gray Slug.‡
s, Shell ; *t*, tentacles ; *e*, eyes ; *b*, breathing-hole.

Yet, even kept down as they are, there is not vegetable food enough for all kinds, and many feed on

* Helix. † Testacella. ‡ Limax.

other animals, as for example the little testacella (B, Fig. 46), a queer little fellow which follows the worms down into their holes, and drags them down his throat by his rasp of barbed teeth, so that often several worms may be found torn and mangled within his body. His breathing chamber has found its way nearly to the end of his tail, so that he can breathe when the front of his body is buried, while the little shell (*s*) which covers it looks very comical, but is useful, nevertheless, in protecting it from attack behind.

All these many forms of **water-snails**, and **land-snails**, and **slugs**, have taken possession of the land and its waters, and now if we go back to the sea we find that the world has still room for other kinds, only they must fit into gaps that are not occupied. For wonderfully beautiful mantle-covered creatures may be found there lurking under stones and in dark corners, if a careful search is made at low tide. These are commonly called "sea-slugs," and by scientific men the "naked-gilled" mollusca, because they have no shell or covering over their feather-like gills (*g g*, Fig. 47), but carry them erect on their backs like tufts of moss or delicate seaweed. Yet in their babyhood these naked animals lived in a tiny curled shell, and swam about by lashes like the young of all the stomach-footed animals, and we can still recognise their nationality, by their feathery gills and their coiled rasping tongue. Like the land-slugs they can creep through many a narrow opening not possible for shelled animals, and though their eyes are not powerful they have very sharp ears, a quick sense of touch, and sensitive nerves. Especially their smell is very acute, probably in order to prevent them from venturing into bad water where their delicate and unprotected gills would be unable to work well. Though they are so fragile-looking, yet they eat ravenously, feeding on young corals, sertularias, and sponges, and often digging a good piece of flesh out of a sea-anemone with their scoop-like rasp. Some

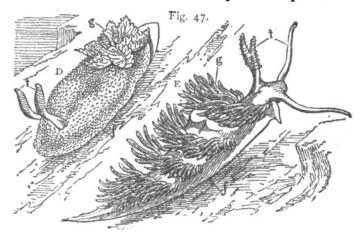

Naked-gilled Mollusca, commonly called sea-slugs.—*Alder and Hancock.*
D, *Doris pilosa.*   E, *Eolis coronata.*   *f,* Foot ; *g,* breathing-gills ; *t,* tentacles.

of them are protected by spicules set in their flesh, but most of them are very tender, and escape observation by the wonderful resemblance of their colours to those of the seaweed over which they wander ; and whether floating, or hanging by slimy threads, or crawling with their beautiful plumes outspread, they select chiefly the dark sheltered spots neglected by the hardier children of Life.

THE MANTLE-COVERED ANIMALS. 125

And now that the sea-shore, the ponds, and the rivers are overrun with stomach-footed animals, there remains but the wide ocean. And even there they have made their way, for sailors in the Atlantic Ocean meet with the ocean-snail (Ianthina), with its float of air-cells, floating in myriads over the sea and feeding on the small jelly-fish, and with the lovely

Fig. 48.

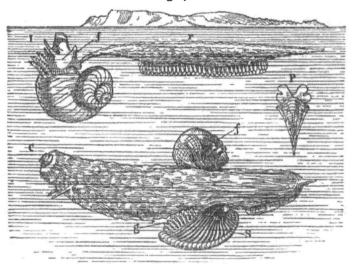

Oceanic Mollusca.

I, Ianthina, the ocean-snail. *f*, Foot; *r*, raft of air-bubbles, with egg-bags hanging down. C, Carinaria.* *f*, Foot; *s*, shell covering the breathing-gills, *g*, both these forms float upside down. P, Pteropod or wing-footed snail.

Carinaria, whose foot has been moulded into fins (*f*, C, Fig. 48) with which it swims upside down in the water, its delicate shell serving to protect its breath-

* *Carinaria atlantica.*

ing-gills (*g*). And as the Carinaria swims along he feeds on other and minute univalve animals, such as the sea-nymphs and wing-footed snails (*Pteropods*), which discolour the water for miles with their swarms, as they graze on the floating seaweed.

Life then has spread her mantle-covered children far and wide over sea and land, where each by different devices finds food and shelter. But it is not with such tiny beings as these that we are to end the history of the mantle-covered animals; for lurking in the holes and tide-pools of the sea, there are much larger creatures with sac-like bodies, green staring eyes, horny beaks, and waving arms, which, unlike as they are to the ordinary shell-animals, are nevertheless true mantle-bearers.

Who would imagine, on seeing a cuttle-fish with its large pathetic eyes, thrown up on the sea-shore, or an octopus shooting across its tank, that these intelligent, active creatures had any connection with the helpless oyster or timid periwinkle? Yet so it is; only while the oyster is one of the lower and feebler forms, the cuttle-fish, the octopus, the argonaut, and the nautilus, are the monarchs of the mollusca, provided with as powerful weapons for their work as the dragon-fly is among insects or the tiger among beasts.

Go some day and look at an octopus in one of the aquariums. Its bag-like body appears to be a mere mass of flesh; yet it has really a most complicated internal structure, and a gristly framework more like a true skeleton than any other animal without a backbone. Its mantle covers the body and forms a ring round the neck, often fitting so closely that its edge can only be seen where there is a hole for taking in water. In a fold of this mantle are hidden the gills, and a short funnel (*si*, Fig. 49) sticking out of its neck is a tube for shooting out the refuse water which has been taken in at the mantle-rim. Here we have the secret of the rapid movements of the octopus, for, by taking in a supply of water at the rim of his mantle and sending it out in jets through the funnel, he shoots

Fig. 49.

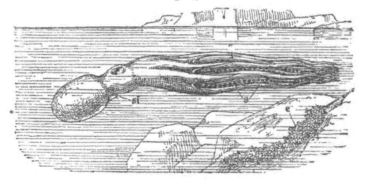

Octopus shooting backwards through the water.

si, Siphon; *a*, arms; *s*, suckers on the arms; *e*, a bunch of eggs of the octopus.

himself backwards just as a boat is sent through the water by a stroke of the oar. Nay, more, if he is flying away from an enemy he has an additional mode of defence, for within his body is a gland which secretes an inky fluid, and this he squirts out through the funnel, making a thick dark cloud behind him which baffles his pursuer at the same time that it helps himself to dart away.

> "Th' endangered mollusk thus evades his fears,
> And native hoards of fluid safety wears.
> A pitchy ink peculiar glands supply,
> Whose shades the sharpest beam of light defy.
> Pursued he bids the sable fountain flow,
> And wrapt in clouds eludes the impending foe."

Fishermen assert, and Mr. Darwin and others confirm their opinion, that the octopus and cuttle-fish often take deliberate aim at an enemy when they squirt out this unpleasant fountain.

But the chief and most powerful weapons of the octopus are his so-called arms and his horny beak. Just below his large penetrating eyes is spread out a crown of eight long tapering ribands (*a*, Fig. 49), and these are, in fact, his foot, answering to that crumpled muscular disk upon which the snail walks. In the octopus this foot has grown round the neck and then divided up into segments, and for this reason he and the cuttle-fish and nautilus are called *head-footed* animals (*Cephalopoda*). The foot of the cuttles has ten segments instead of eight, and two are nearly three times as long as the others.

Now watch the octopus lurking in the rockwork of the tank, his round body squeezed into some nook, and his arms,[*] some grasping the rock, others flapping idly in the water. If a large fish or crab pass by instantly he is on the alert; the arms in the water, no longer listless, dart out and fasten on the luckless animal, which is dragged in to the strong beak standing out in the centre of the arms and crunched in a moment, even the crab's shell cracking like a nut, while his flesh is devoured and carried down into the

---

[*] For so we must call them, although they are really strips of his foot.

stomach of the octopus by his fleshy tongue armed with horny hooks. But what gives the arms of the octopus such power? If you look at the under surface of them you will find, arranged in pairs along each arm, suckers (*s*, Fig. 49), large near the mouth and growing small as the strips taper to a point, and crowded so thickly that an ordinary-sized octopus with arms about a foot and a half long will have nearly 2000 of them. Each of these suckers is a perfect little air-pump with a piston in the middle, and the moment the octopus lays an arm upon any creature, a muscle draws the piston in each sucker back. This causes it to cling like a cupping-glass, and the more the victim struggles the tighter is the grasp; while the octopus holding by the suckers of his other arms to the rock has a firmer and firmer hold the stronger the resistance.

One would almost imagine at first sight that long experience would have taught the fishes and crabs to keep out of the way of such a monster; but the octopus has another, and almost unfair, advantage. He carries in his transparent skin cells of colour, yellow, blue, red, and brown, and has the power, like the chameleon, of changing colour and assuming the tint of the rock under which he hides.

> "New forms they take, and wear a borrowed dress,
> Mock the true stone, and colours well express.
> As the rock looks they take a different stain—
> Dappled with gray, or blanch the livid vein."

By this means he not only lies safely in wait to pounce upon his prey, but may himself escape the notice of the dolphins or the conger eels, which are too strong for him to conquer, and who in their turn feed on his fleshy arms.

With such advantages and weapons of attack, can we wonder that not only the octopus but also his ten-armed relations, the cuttles and the squids, are to be found of different sizes and kinds all over the sea? There is the little *Sepiola*, often caught off our coasts in the nets of the shrimpers, whose body is only about half an inch long, with small flaps or fins on the sides. He, like the cuttle-fish, so far clings to the old habits of the mollusca as to form a long thin shell on his back *under* his mantle; and this shell we call a "pen" when we find it on the shore because it is shaped like one. He makes himself a shelter by blowing a hole in the sand with jets of water from his funnel, and uses the suckers of his arms to remove and arrange the small stones. Then he sits in his hole, with his large goggle eyes peering out, and catches the shrimps and smaller crabs as they pass by. There is the common cuttle-fish which forms in its mantle the white chalky shell known as the "cuttle-bone." It generally floats about or creeps over the bottom of rocky pools; till frightened, or, wishing to attack some animal, it shoots out suddenly a jet from its funnel and flies backwards through the water, clutching its prey on the road. The dark horny grape-like bunches which we find on the shore are the eggs of the cuttle-fish. There are the Calamaries, whose shell is a horny "pen," and some of which living in the open ocean have sharp hooks in the centre of their suckers, making cruel weapons of attack against the unfortunate fish, who have the sharp hooks planted in their flesh and held fast by the cups around them.

Then there is our friend the Octopus with his body squeezed between the rocks and nothing but his bright, gleaming eyes to betray him, while his wife in another sheltered nook is watching over her eggs (*e*, Fig. 49) arranged in clusters on a stalk like a huge catkin of a nut-tree. A loving mother she is, sometimes dandling the eggs in the hollow web of her arms or cleaning them by spouting water from her funnel over them, as a gardener washes his plants with a hose. Week after week she will watch them, for though they do not need hatching, yet if she did not keep them clean they would be addled by living things growing over them; then as each little bag bursts a tiny perfect octopus about the size of a flea darts out, uses his funnel at once, and frolics to and fro in the water, his body blushing now with one colour and now with another.

In our seas an octopus scarcely ever has arms more than two feet long, and a body about the size of an ordinary lemon; but in the Mediterranean they have been caught with arms four feet long and are much dreaded by the bathers, and in the British Museum there is an arm of a Calamary nine feet in length, so that the creature which carried it and which probably lived on the coasts of South America, must have been formidable indeed.

But if there are ugly and dangerous "head-footed" animals, there are among them two lovely forms. The Argonaut, though she does not really sail on the water with her two arms raised as sails, as the poets imagined, yet forms such a lovely cradle for her eggs, which she carries with her, that it makes her a "thing of beauty" as she drives herself backwards through the water. The shell-bearing Argonaut is the mother, for the father is like an ordinary octopus and has no covering; and indeed that which the mother carries

Fig. 50

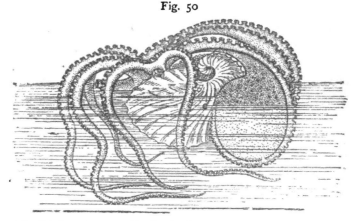

The Mother Argonaut floating in the water.*—*Verany.* *e*, Eggs.

is not a true shell, but a chalky nest built up by the ends of two of her arms, which are spread out into broad webs and folded back over her body where they lay down that beautiful delicate film of lime, the "Argonaut shell." Under this shell, still keeping it covered with her arms, she places her bunches of eggs, and stretching out the other six arms, can fly backwards through the water carrying her brood with her, or can, like the cuttle-fish, float quietly or creep along the bottom.

But perhaps the most beautiful shell of all is that

* When in rapid movement the arms are in a straight line, as in the Octopus (Fig. 49).

of the Nautilus, which, it must be remembered, is totally different from the Argonaut shell, being the animal's real home and not a mere nest. The Nautilus is different in many ways from the octopus and the cuttle. He has four breathing gills instead of two; his eyes are much less perfect than those of the other head-footed animals; he has no ink-bag, for having a strong protecting shell he has less need for it; and he has no suckers on his feet. He is the last remnant of a once great family, that of the huge Ammonites and Nautiluses, which we find buried in the rocks of ages past; and, like many a remnant of a once noble race, living retired in their own domain while younger and less sensitive branches are fighting their way to eminence in the world, the nautilus creeps in the shallow waters of the Indian Ocean, or floats about when all is still, but hides himself persistently from view, and has very rarely been seen alive. He builds his beautiful and refined house chamber by chamber, deserting one after another as he grows too large for them, and leaving only a thin tube through the middle, by which he is supposed to fill the shell with air when he wishes to float.

> "Year after year beheld the silent toil
> That spread his lustrous coil;
> Still, as the spiral grew
> He left the past year's dwelling for the new,
> Stole with soft step its shining archway through,
> Built up its idle door,
> Stretched in his last-found home, and knew the old no more."

He allows very few, however, to investigate his habits: wrapped in his proud reserve he lives his solitary life, and it is only after his death that his beautiful shell with its pearly chambers is found and brought to decorate our homes.

And here we must take leave of the mantle-covered animals. We have followed them, though very imperfectly, from the "poor patient oyster," through their gradual rise in power ; till we leave them as dreaded conquerors, in the sharp-beaked octopus and the terribly armed calamary. We might, if we had ventured on the dangerous sea of conjecture, have started still earlier, and linked their simpler forms to those of the lower worms. But till more is known, this course might have led us astray, and it is safer to content ourselves with marking how life has gradually filled the ocean and the land with specially fitted forms of mollusca, having all a distinctive nationality which separates them from the other divisions of Life's children ; so that the octopus, the cuttle-fish, and the nautilus, stand as undoubtedly at the head of one great plan of animal life, as the ants do at the head of the insects, or man at the head of the vertebrates. We shall now have to hark back again, and in inquiring of the worm whence he comes, and how he lives, start on a totally different track, which will lead both by land and water, through the forms of the shrimp, and crab, and lobster, to the aërial and fairy-like insects which form so large a portion of the life upon our globe.

INSECT LIFE.
(for description see list of illustrations.)

CHAPTER VII.

THE OUTCASTS OF ANIMAL LIFE, AND THE ELASTIC-RINGED ANIMALS BY SEA AND BY LAND.

> " And ever at the loom of Birth
> The Mighty Mother weaves and sings;
> She weaves—fresh robes for mangled earth;
> She sings—fresh hopes for desperate things."
> KINGSLEY.

WE have now traced the history of four out of the seven divisions of animal life, and have seen how each, by taking a different road, has managed to get a footing for its members in various nooks and spaces in the world. We must next try to gain some idea of that small fifth division containing the Worms; in which is shadowed forth, as it were, that ringed structure which we shall find so remarkable in the sixth and largest division which follows. But, before arriving at the true ringed worms, we must pause for a moment to glance at that curious, wandering, and outcast population of our globe, which, finding no shelter in the earth, or sea, or air, have taken up their abode within their fellow-creatures and live upon them.

Although we have as yet studied only the lowest, and by no means the most numerous of Life's children, yet we begin to see that our earth is full, very full, of life, and that the creatures in it are jostling each other, and driving into dark and dismal corners those which cannot get a living in the open sunshine. Millions serve as food for others, and millions die a speedy death from want of space and food; but we cannot expect that any will give up their lives while they can find a means of struggling on. What way is there, beyond those which we have found already?

There is still the novel device of a creature finding shelter by making another living being carry it, and of obtaining food by making another living being nourish it. And so we find that among the low forms of many classes of animals there are always some which prey upon their neighbours, just as in our great cities there are always some of the most degraded and miserable—our street Arabs and our thieves—who live on refuse and plunder.

And this is true to such a large extent in the animal world, that there is probably scarcely a single creature that does not carry many other creatures upon or within its body.

Some of these merely come to it for shelter, as, for example, the tiny pea-crab, which is constantly found living in the shell of the horse-mussel, catching its own food, and being probably rather helpful than otherwise to the mussel, by leaving him the scraps of his meal. Others, such as ticks and water-mites, fix themselves on the bodies, the one of sheep and dogs, the other of water-beetles, and sucking the blood of their hosts, find both food and shelter. And others, finding no place for them at all in the outer world, burrow into the very body of their victim, and feed upon the soft parts within.

Among these last, the greater number are a low race of soft-bodied worms, whose ancestors, when the other forms of life—the star-fish, mollusca, ringed animals, and insects—found new ways of gaining their livelihood, remained behind, groping in the mud and sand of rivers and seas, and flapping about by the broad margins of their flat bodies. Some of the descendants of these soft-bodied worms still manage to live a free and independent life. One set called the wheel-worms,* because of the curious whirling appearance of their lashes as they swim about, may be seen under the microscope in almost any stagnant water. Another group, with tiny red eye-specks, and a trumpet-shaped mouth in the middle of their bodies,† live on the sea-shore or in ditches, and may be found as little jelly-lumps upon water-cresses before they have been washed. Another set,‡ known as the "ribbon-worms," with elastic bodies which stretch sometimes to an enormous length, are armed with a tiny dagger in the head, with which they pierce the soft bodies of animals and suck out their juices. One of these called the long-worm,§ which looks like a dark strip of india-rubber as it lies coiled up under stones on the shore, has been known to be as much as twenty feet long, though only as broad as the blade of a pen-knife.

These are the more fortunate of the soft-worms which have found a place in the outside world; but

* Rotifera.   † Planaria.   ‡ Nemerteans.   § *Nemertes borlasia.*

there are others which, unable to get a living in the mud and sand, were forced to work their way into the bodies of snails, caterpillars, or grubs, and now make them their natural home. Unpleasant as it may be to think of these parasites, yet when we look at the question from their point of view, they are after all only doing their best to get a living, and they have many curious weapons to help them in doing it, nor do they always injure the animal upon which they live, unless they are in great numbers.

Thus, for example, one of the flukes,* a minute flat-worm shaped like a tiny flounder, has a most strange succession of changes in its life. Firstly, The mother lives within the intestines of some water-bird, holding on firmly to her host by two rows of tiny hooks round her head, while her mouth is firmly applied like a sucker; secondly, the eggs are thrown out and fall into the water or moist mud, and out of them comes, thirdly, the embryo or imperfect animal, surrounded with lashes; but it does not long remain free, for out of it again comes a fourth form, a small bag-like animal, which at once seeks out a water-snail (*Paludina*) and clings to it. Nor are the transformations yet ended. Within this hanging sac, which is called the "nurse" of the fluke, there appear, fifthly, a number of little tailed animals like tadpoles, and by and by the nurse bursts, and all these little creatures come swimming out once more free in the water. But the snail is not rid of them; either upon her or upon some other snail like her, a number of these little creatures fix

* *Distoma militare.*

themselves, and each one boring into her foot, drops off its tail, and forming a transparent bag round its body, begins to grow a crown of hooklets. In this state it remains till the snail, gobbled up by some water-bird, passes into its stomach, and there the gastric juice, digesting the snail, dissolves the bag, and at last the fluke becomes a perfect animal again, fixing itself by hooks and suckers in the same kind of home from which its mother came.

And now consider what a number of chances occur to this animal during its short life, any of which may destroy it. Their eggs are not placed in a fit spot by a careful mother, but fall wherever the bird may chance to drop them, and twice in their lives they have to find a snail in which alone they can live and grow. Many fail, and clinging to stones or weeds, die for want of their home. And even if they succeed in these first attempts, the last step of all is entirely out of their control, for unless they are carried down the throat of the water-bird, they can never grow and lay eggs. But they exist in such myriads that this is of no consequence to the race. You can scarcely cut open any snail without finding some of these curious creatures within it, different species living in different snails; and in most cases the worm must pass into another animal to become complete. The liver-fluke of the sheep for example, which causes the "rot" when too abundant, lives its early life in a snail, which is licked up by the sheep as they eat the damp grass.

The bladder-worm, however, which gets into the brain of the sheep, and causes it to hang its head, belongs to another and perhaps more dangerous tribe. These are the so-called "tape-worms" which can only grow to their full strength in warm-blooded animals, and are armed with both hooks and suckers on the head. Now, while the front part of this head is firmly fixed, buds are given off continually from the other end, making a long tail with many joints, each of which carries eggs, and often has its own separate suckers and hooks to hold firmly to its host. These creatures have no mouths or stomachs, but take in the fluid food all over their body as it passes by them on its way through the animal they inhabit. Tape-worms wander just as flukes do, thus the tape-worm of the dog begins its life in the sheep, that of the cat lives first in the mouse, that of the fox in the hare or rabbit, that of the water-bird in the fish.

Nor is it only flat-worms which have become parasites; the little wriggling round worms live, many of them, in the grubs of beetles and insects, and from these pass on into the bodies of rats and mice, squirrels and birds, or fishes. The little thread-worm *Mermis*, for example, as soon as it is hatched in the moist earth in spring time, uses a sharp dagger hidden in its head to pierce a road for itself into the body of a grub, and lives upon its juices till either the caterpillar becomes a butterfly, or is eaten, or the mermis is ready to lay her eggs, and then she pierces her way out again to lay her young in the soft earth.

Another little round worm hangs on by its suckers inside the throat of the chicken, giving it the "gapes," which can be cured if the worm is brushed out with a feather; while the *Trichina* so dangerous in half-raw pork or ham, is another round worm, living in the muscles of the pig. All these, and hundreds of forms like them, belong to that wandering band of outcasts, which have been driven from the face of the earth to feed upon the strength of others. They are not a pleasant band, but they teach us most surely the truth that the children of Life are sown broadcast over the earth, to make the utmost use of it that can be made. We have even examples where a parasite upon some animal has another parasite within it; as when by cutting open a snail, worms are found within, and these worms when cut open are found to be the home of some tiny infusorian or slime animal, so that even within the body of one animal we have a little world of life.

Another truth it teaches us which we have noticed before; namely, that where a creature has little use for its powers, these diminish and it becomes degraded and feeble; for the parasitic worms, with their low structure, their want of eyes and ears, and often of mouths and stomachs, are most of them poor miserable creatures at best. Yet still we find even here that each must do some work. The most shiftless of worms passed on passively from one animal to another, must find its way to the liver, or the muscle, or the intestine which is its natural home; and in the hooks and suckers, and daggers so admirably fitted for opening a path, and clinging firmly when the right spot is found, we see a proof that even these poor debased parasites have acquired some weapons in the struggle for life.

But we must not stop here in our history of the worm tribe, for these parasites have distant relations of a far higher structure, who have managed to gain a much better position in the world. In each of our groups of animal life we have found some special advantage which has enabled them to spread their children over the world; the sponges had their co-operative life and their protecting skeletons, the lasso-throwers their poisonous weapons, the prickly-skinned animals their tube feet and stony casing, the mollusca their wonder-working mantle, but among them all we have not yet met with that power of moving quickly, without which no creature is ever very intelligent. It is true that the octopus can shoot rapidly through the water, and is at the same time the most intelligent animal we have yet learned to know; but its quick movements are all in the water; when it scrambles along the shore it is slow and awkward, while the other crawlers, the sluggish snail or the creeping star-fish, are not any more rapid. And yet it is clear that the power of getting quickly over the ground must be an advantage in the struggle for life, and we shall see that it is this power and the intelligence accompanying it which has raised the most advanced animals in the sixth division to such a high position as that of the bee and the ant.

Nothing, however, is learnt in a moment, and therefore you must not be surprised that the worm and the leech, which you would probably consider rather slow animals, are the first examples of the more active creatures. Nevertheless, if you could start either of these animals on a fair race with a snail, though they might not appear to hurry yet you would find they would beat him hollow. The accompanying picture is one given by Sir Emerson

THE OUTCASTS OF ANIMAL LIFE. 143

Tennent of the land-leeches as he saw them in the low ranges of the hill country of Ceylon. He tells us that these little leeches, about an inch long, fixing themselves by their tail suckers, raise their heads in the grass to watch for passers by, and as soon as they see man or beast they start off. Now stretched out at full length, now drawing up the hind sucker so as to form a loop, then forward again, they advance at an astonishing pace till they reach their

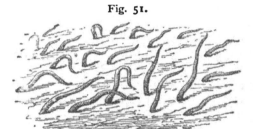

Fig. 51.

Land-Leeches of Ceylon* racing to attack some creature.

victim, when they cling to ankle or leg, or even if these are protected are soon up at the neck, where they hang in groups like bunches of grapes, as their skins swell out with their meal.

Now, if we wish to learn the secret of the leech and how he can move so fast, we must look for it in two things. 1st, in the muscles by means of which he moves his ringed body ; and 2dly, in the chain of nerves which give the order for the muscles to move. He has three layers of muscle in his skin—in the first, nearest the outside, the fibres run round and round the body in rings, in the second they cross each

* *Hæmadipsa Ceylonica*, Sir E. Tennent, *Ceylon*, vol. ii.

other making a diamond-shaped lattice-work like a netted purse, in the third they run along the body from head to foot. When the leech wishes to lengthen his body he contracts the round rings and so forces the long cords to stretch, making himself long and thin; when he wishes to shorten his body he contracts the long cords and forces out the rings, making himself short and stout, while the criss-cross muscles help to modify these movements.

So much for the muscles, and now for the telegraph which governs them. If you were to lay a dead leech on its back and open it, you would see running from end to end of its body a white cord (c) with little swellings of white matter (g) at intervals upon it, and from these swellings very fine white threads (n) are seen branching out into the body. The cord is made of nerve-threads clinging closely together, and is so to speak the line of telegraph; the swellings are masses of nervous matter called *ganglia*, and are the telegraphic stations; the white threads are simple nerves carrying messages to the muscles; while round the neck of the leech is a collar

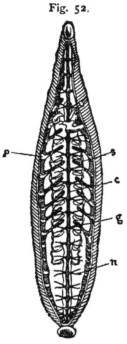

Fig. 52.

After Moquin-Tandon.
Section of a Leech,* to show the nerve-cord c, with the ganglia or knots of nerve-matter g, and the nerves n, branching off from them; s, walls of segments of the body; p, pockets of slime.

* *Hirudo medicinalis.*

of nerves with two large *ganglia*, the head telegraph offices. Now, it is this system of nerves which enables the leech to give orders to its muscles so rapidly, and throughout all the ringed animals this same system is found growing more and more perfect up to the ants.

When the leech is alive and uninjured, all the telegraphic stations work together, and you will notice that in the middle of the body, which is divided into segments (*s*), each has its own station or ganglion, and though all these usually work together, yet each segment is so active that if the cord is cut in half in the middle, the stations in the tail end of the leech will work on their own account and the two halves will often try to pull different ways. We see then that we have here a very powerful machine, and when we remember that the leech has eight or ten simple eyes set in its back near the head, and two strong suckers to cling with, within one of which is a mouth armed with three saw-like jaws which can easily pierce the skin of its victim, already made tight by the sucker, we can understand that he is well fitted for the battle of life. He is essentially an aquatic-breathing animal; and though he can live for some time out of the water, he can only do so in very damp air, and his body is always covered with slime which oozes out from some little round pockets (*p*) in the sides of his body.

So the leeches live in ponds, and ditches, and marshes, and some even on damp land; and the eggs out of which the young leeches come, are laid in cocoons of gummy slime placed in the holes and clay of the banks. Fish, snails, limpets, and grubs are their usual food, though they by no means despise warm-blooded animals when they get a chance to fasten upon them.

The elastic-ringed animals are not, however, confined to fresh water; on the contrary, though they cannot breathe in perfectly dry air, yet they have found their way underground in the common earthworm, and there are many of them in the sea, from which probably they first came, and where they are protected and armed in many very curious ways.

The common earthworm, which we all know so well, is a curious example of a water-animal adapted to live under the earth. He breathes as the leech does, and he must have moisture, for perfectly dry air is useless to him, and he dies quickly in very dry places where he cannot keep his body moistened with slime. Eyes would be of no use in his underground journeys, and he only comes above ground at night, so we find that these organs are wanting; suckers too would be a hindrance to him, and his body ends in a fine tapering point which he can push into the earth like a shoemaker's awl.

But how is he to force his way through the earth? If you pass your hand along his body from the tail to the head you will feel a gentle resistance, for every ring bears four pair of hooked bristles pointing backwards, so fine as not to be easily seen, but strong enough for his work. When he has pushed the front part of his body a little way into the earth he then draws it up by shortening the long muscles, and the bristles make no resistance because they point towards the tail; then he contracts his ring muscles and so forces his body to lengthen again, but this time it cannot lengthen backwards, because the bristles being rubbed the wrong way will not yield, but stick into the earth, so that the whole movement is forwards, and he makes his way.

He often assists himself too in another way by eating the earth through which he passes; he has no hard jaws like the leech, but a long upper lip with which he shovels the earth into himself, sending it out afterwards at his tail, and making those curious coils of earth which we find on lawns and garden paths. His usual food is the animal and vegetable matter in the earth, which he absorbs out of it as it passes through his body, though it is possible he may also sometimes eat the leaves which he is so fond of dragging with him underground, leaving the stalks sticking out above. The young earthworms are hatched underground in cocoons made of earthy matter and slime, and as they have no eyes or tentacles or other tender organs, they become at once fearless miners. Yet they often fall victims at all ages to the hedgehog and the mole, and even to their relations the leeches if they venture near the water; while birds are their mortal enemies. Even if a bird cannot succeed in catching a whole worm, yet he will often nip off his tail as he is disappearing into the earth in the early morning after his nightly rambles. As, however, the worm can grow the tail again without any difficulty, the loss is perhaps not of much consequence; and from his living underground he is certainly exposed to fewer dangers than our next examples, the seaworms, which are obliged to protect themselves in many ingenious ways.

Very few people, as a rule, are acquainted with the seaworms in their homes, but every one who has handled oysters or scallops must have noticed the curious round tubes often firmly clinging to their shells. These tubes were once the home of a sea-worm which has built them of chalk and slime. The worm itself is quite loose within the tube and stretches

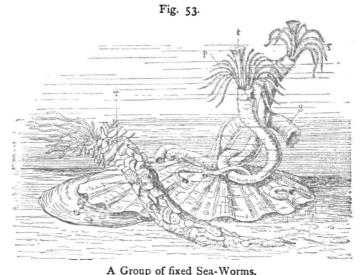

Fig. 53.

A Group of fixed Sea-Worms.
S, *Serpula*.   t, Tentacle; *p*, breathing-plume.   T, *Terebella*.
Sp, *Spirorbis*.

its body out, scrambling up the sides by the help of its bristles and of a number of little comb-like plates which grow upon its rings. If you can get a shell covered with these tubes from the sea at low tide and put it in salt water, you will see a beautiful sight. After a time a small scarlet stopper (*t*, Fig. 53) will creep up and out of the tube, and as it rises on a long stem there will follow it a splendid scarlet plume (*p*) arranged like a double fan, and waving in the water. The stopper with its stem is one of the tentacles of the worm enlarged at its end so as to shut the animal safely within the tube, while the other tentacles have become the beautiful plume which is the breathing apparatus of the animal. It is easy to understand that being in a tube, the Serpula, as this worm is called, cannot breathe through its skin like the leech or worm, and it needs these delicate gills to provide air for its body, while at the same time its sensitive nerves and apparatus of muscles enable it to draw them in like lightning when danger is near.

There is an almost endless variety of these tube-building worms. You can scarcely pick up a piece of dark seaweed without finding upon it what look like very tiny shells (Sp, Fig. 53), but which are really coiled worm-tubes. Again you cannot search long among the sandy pools at low tide without finding some long tubes made of sand and broken pieces of shell wedged between the stones and rocks, and having forked sandy threads at their end. These tubes are the house of the Terebella or shell-binding worm, which selects particles of shell and sand with its tentacles and places them round its soft body, cementing them together as a mason cements the stones of a wall, till it forms a tube often a foot long, so firmly wedged into the beach that it is almost impossible to get one out perfect; while you will rarely find the worm itself, as it draws back to the farthest end of the tube directly it is alarmed.

These are the fixed seaworms, but there are

others as active as any animal in the sea, and the first step towards these is the common *lugworm* which fishermen use for bait. This worm, which makes the round coils of sand we meet with on the coast, moves freely about but is not very active, for it has no eyes and lives much underground, glueing together the sand as it passes along, and forming a tunnel for itself through which it can pass. Its gills are no longer round its head, as among the fixed worms, but it carries them on its back as thirteen pairs of lovely scarlet tufts.

And now we come to the wonderful defensive weapons which life has bestowed upon these wandering worms. The lugworm safely hidden in its tunnel does not need any, but the lovely Nereis (N, Fig. 54), which has a well-developed head, with eyes, tentacles, and sharp jaws, leads a much more active and precarious life. It hides under stones and shells, or moves about rapidly in the water, and can use its bristles not only as oars to swim with but also as swords, sabres, and hooks. For these fine bristles are not simple hairs as they appear, but have saw-like edges and hooked tips, and are really formidable weapons, both of attack and defence, although the smaller specimens of the creature which you find on the shore often look like mere threads, unless seen under a magnifying glass.

But if the Nereis is beautiful and terrible, how much more so is the marvellous sea-mouse (A, Fig. 54), which we sometimes see thrown up on the shore, while small ones may be found by turning up stones on the sand. No one would believe at first sight that this creature is a worm, covered as it is with broad scales and bristling with tufts of hair. Yet if

ELASTIC-RINGED ANIMALS.

you lift the scales and brush aside a thick coating of felt which covers the body, or if you look underneath the creature as it crawls along, you will be able to distinguish the rings, and also to see that the tufts of hair spring each from a separate ring like the hairs of the earthworm. The broad scales are a curious breathing arrangement peculiar to the sea-mouse, for when they

Fig. 54.

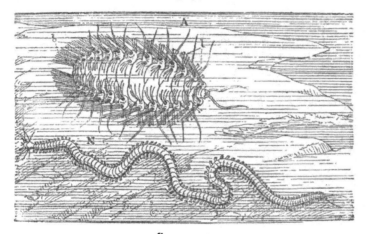

Sea-worms.

A, *Aphrodite aculeata*, commonly called a sea-mouse. *b*, Bristles; *l*, lances. N, *Nereis pelagica*.

are lifted the thick coating of felt is filled with water, which in this way flows all over the outside of the body; and when they are shut they force it out again, making room for a fresh supply of water to pour in when they rise. In this way the whole body of the animal is bathed in water, out of which the oxygen can be taken through the delicate skin.

But it is above all the tufts of hair which are so

beautiful and wonderful—beautiful because each bristle, being marked with the finest possible scratches, reflects light of all the colours of the rainbow—crimson, scarlet, orange, yellow, green, blue, and lilac—according to the angle at which the light falls upon it, so that the creature looks as if it carried a forest of prisms upon its back. Wonderful, because each of these hairs is a sharp lance, by which the worm can protect itself from attack. In one of the sea-mice, the *Aphrodite hispida*, these bristles are perfect harpoons, with barbed points at their tip and delicate teeth all along the edges, and they can be thrust out when the animal wishes to defend itself. But how, then, can the worm avoid cutting itself with these sharp instruments? To prevent this each barbed spine has a smooth horny sheath, which closes upon it as it is drawn in and prevents it from tearing the tender flesh! Such a creature as this deserves indeed to be called the king of worms, being at the same time so beautiful and so formidably armed. He lives in deep water, and is only to be found when thrown on shore, where he is very helpless, though in his own element he is a dangerous neighbour, as he feeds greedily upon all living animals, not sparing even his own brothers when they are weaker than himself. He is a timid creature, hiding under stones and in dark corners and shunning the light of day which gives him all his beauty, yet, in bidding adieu to the worm-tribe, we must acknowledge that none of them can compare, either in delicacy of structure or in their weapons of attack and defence, with the little sea-mouse, or, as he is often called, the " porcupine of the ocean."

CHAPTER VIII.

THE MAILED WARRIORS OF THE SEA WITH RINGED BODIES AND JOINTED FEET.

"Strong suits of armour round their bodies close,
Which, like thick anvils, blunt the force of blows;
In wheeling marches form'd, oblique they go—
With harpy claws their limbs are armed below;
Fell shears the passage to their mouth command,
From out their flesh their bones by nature stand,
Broad spread their backs, their shining shoulders rise;
Unnumber'd joints distort their lengthened thighs;
With stony gloves their hands are firmly cased;
Their round black eyeballs in their bosom placed;
On eight long feet the wondrous warriors tread,
And either end alike appears a head;
These, mortal wits to name as 'Crabs' agree—
The gods have other names for things than we."
"Battle of the Frogs and Mice."

HAVING now arrived at the sixth and largest division of the whole animal kingdom, we are going to leave behind us those low and scattered tribes, which live as it were in a dreamy unconscious way, tossed hither and hither by outward circumstances, and having but feeble nerves to guide them; and for the future shall have to do with beings gradually struggling into active intelligent life.

No one can watch the beautiful transparent prawn, with his bright eyes gleaming, and his antennæ trembling in the water, without feeling that we have here a creature much more alive to everything around him than the groping star-fish or the creeping worm, while the active little crab as he peers out from the seaweed, and scrambles across the shallow pools, or buries himself in an instant in the wet sand, shows a lightness and agility which we look for in vain in the sluggish snail or the slowly-grazing limpet. And when we learn that the prawn and the crab in the sea are formed on the same plan as the centipedes, spiders, and insects of the land, we see that we are on the road to even more intelligent and more active creatures, such as the busy bee and the thrifty ant.

But how can this be, that the heavy armour-covered crab and lobster, which are called *Crustacea* from their hard crust-like shells, should belong to the same type as the delicate hovering butterfly, and the buzzing gnat? Let us pause and master this, for till we have done so, we cannot understand the wonderful way in which the creatures of each group in this division have been adapted to the life they have to lead.

In Fig. 55 we have four animals—a prawn, a centipede, a spider, and a caterpillar together with the butterfly into which it turns. Now all these animals wear their skeleton, or the hard part of their bodies, not inside as we do with soft flesh growing over it, but outside; so that if you grasp any of them when dead, the skin (as we should call it) will bend or crack like a piece of thin horn. Moreover, this hard outside skeleton is arranged more or less in rings with softer skin between them, as you may see in the centipede and caterpillar, and in the hind part of the prawn and butterfly; and they are to be traced in many spiders, though as a rule they have disap-

THE MAILED WARRIORS OF THE SEA. 155

peared. These rings remind us of the worm, only that in the animals of which we are now speaking they are more marked, and whereas the worm has only hairs for legs, these animals have many-jointed limbs which are of great use in running, leaping,

Fig. 55.

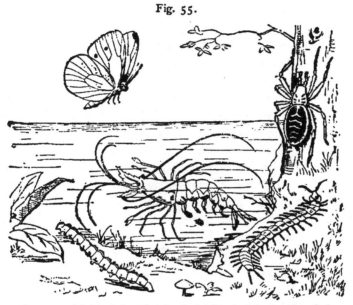

A group of Jointed-footed animals (*Arthropoda*), showing their ringed bodies.

and seizing prey. It is because of these jointed limbs that the crustacea, centipedes, spiders, and insects are all called *Arthropoda*, or jointed-footed animals.* Linnæus called them all *Insects*, because their bodies are cut into divisions (*in* into, *secta* cut), and although naturalists now generally confine the

* *Arthron*, a joint ; *pous*, a foot.

word "insect" to those which have wings and six legs, yet a good English name is so much wanted which will take in centipedes and spiders as well as beetles and butterflies, that I shall follow Mr. Ray Lankester's suggestion* and call all the ringed and jointed-footed animals "Insects."

In this sense the prawns and their relations which are both jointed-footed and cut into parts have been called the "Insects of the Sea," and this name helps to remind us how much they are like the great body of insects on the land.

This likeness is very evident when we compare the four types in Fig. 55. Thus we have first the butterfly, whose body you will notice is cut into three distinct parts—the *head* with one pair of feelers or *antennæ* on its forehead, a pair of eyes on the side of the head, and mouth-jaws below; the *thorax*, or chest, on which grow the six legs and two pair of wings; and the *abdomen*, or hinder part of the body, which never possesses any limbs. The butterfly thus is a six-legged winged insect. Then we have the centipede, whose ringed body reminds us of the caterpillar from which the butterfly springs, but which has jointed feet on every ring. Next we come to the spider, and here we find the head and shoulders joined into one strong piece, and bearing four pair of legs, while the abdomen has nearly lost the traces of rings. The antennæ are bent down over the forehead, and have been turned, as we shall presently see, into pincers, hooks, and poison fangs; while the short feelers in front of the head, which look like

* Haeckel, *History of Creation*, English translation, vol. ii. p. 178, *note*.

antennæ, are really a part of the mouth. Lastly we have the prawn with his head and shoulders joined into one like the spider (in the lobster you may see a curved line marking the spot where these are joined), with five pair of legs, while some of its relations have many more; and the usual ringed abdomen which in this case has little paddles under it for swimming.

Now just as when we feed, part of our food goes to make phosphates, which form and strengthen our bones or internal skeleton, so do all these animals make out of the food in their bodies a substance called *chitine* something like horn, and this is deposited in the outer layer of their skin, and makes a firm skeleton all over the body, and eyes, and antennæ, and legs; and within this firm skeleton the soft animal lives, much as a soldier in olden times was enclosed in his jointed armour. But if a soldier had been placed in armour as a baby, he would have had to change his suit many times before he became a man, and this is also the case with insects. Their covering is not like that of the sea-urchin, which we saw could be added to at every point; it is made once for all, like the soldier's armour, and the creature must throw it off when it becomes too small for its body. Thus the prawn, the centipede, the spider, and the caterpillar alike creep out of their armour many times in their lives, leaving it often standing so perfect that it looks like the creature itself.

We see then that the prawn and his relations, although they live in the sea, belong to the ringed and jointed-footed division, and are formed on the same plan as the land-insects, which have spread so far and wide over the globe. These are an active busy multitude, which, if they could think and speak, would have far more right to call this earth *their world* than we have to call it ours; for whether in the sea, or in the rivers and ponds; in the fields, forests, or marshes; at the tops of mountains, or in underground caves and passages; in our gardens, our cellars, our houses, or about our persons; anywhere, everywhere, all over the world their hosts are to be found.

We are accustomed to attach great importance to the back-boned animals, the fishes, reptiles, birds, lions, elephants, and monkeys, because they are comparatively large and conspicuous, but in truth, if we except the human race, they are as nothing, either in number or in activity and ingenuity, as compared with the insects and their allies.

If we could take one of each species of all the back-boned animals, and add to them all the species of worms, mollusca, prickly-skinned animals, lasso-throwers, sponges, and lime and flint builders, all these together would only make up 50,000 species, or *one-fifth* of the animals on the globe; the other *four-fifths, or 200,000 species, belong to the ringed and jointed-footed animals*, and of these 150,000 are the six-legged insects. Now we have learnt that if creatures succeed in the battle of life, it is because they can hold their own and fight bravely, and therefore we are prepared to find that life has taught these, her active children, many new lessons and armed them with many useful tools and weapons, differing greatly according to the lives they have to lead.

And first of all we must turn our attention to the

great group of "Crustacea," the "insects of the sea." For though some of this group, as the water-flea and cray-fish, live in rivers and ponds, while a few, such as the wood-louse and even some kinds of crabs, crawl upon the land, yet the chief home of the crustacea is the ocean, where, having scarcely any enemies so powerful as themselves except their own relations, they run riot both as to numbers and size. Think for a moment of the multitudes of sandhoppers to be seen leaping on a dry sandy shore in the evening, or which rise like a cloud of dust out of the half-rotten seaweed if you stir it with your hand. Try to reckon up the myriads of shrimps and prawns which must be caught daily to supply all England, and which are nothing to those that remain behind. Look at the large crabs and lobsters in the fishmongers' shops, and think that in London alone 25,000 lobsters are often sold in the season in one single day! Then call to mind how you cannot walk a step on the shore at low tide, without seeing some tiny crab scuttling along in a hurry to catch something, or to escape being caught himself; or how constantly you come across a hermit crab with a periwinkle or whelk shell on his back, making tracks in the sand as he wanders along. Try and count some day the number of acorn shells (Fig. 61, p. 174) which grow on one single piece of rock or the groyne of a pier. For these too are crustaceans, as are also the barnacles (Fig. 61) which hang from floating timber or gather on the bottom of ships. When you have gained some idea of the multitudes of these creatures on our own shores, you will not have reckoned one millionth part of the crustaceans which live in the sea, for not only are there strange forms of all kinds on distant shores, but there are oceanic crabs which swim in the open sea for days without resting, just as the albatross flies over it, while smaller crustaceans swarm under the ice in the Arctic regions, and there is scarcely a fish which has not an animal of this class living on some part of its body.

Of all the many forms, however, there is probably not one more beautiful than the delicate transparent prawn as he paddles along lazily in sea pools, or through the still water of an aquarium. His horny skeleton is so clear and glass-like, that it looks like crystal, while the formidable toothed saw protecting his head, is scarcely visible in the water, and his delicate antennæ and tapering limbs look as if they would snap at a touch. As he swims you will notice that it is not his ten true legs in the front part of his body which row him along, but the little hairy swimmerets, S, which lie under the hinder part or abdomen, while if anything alarms him, he darts rapidly backwards by a smart stroke of his fan-like tail, t. His long antennæ or feelers, a, are streaming over his back, while a pair of shorter antennæ or *antennules* (a^2) as these are called, each

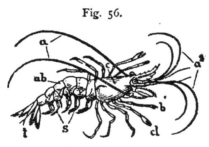

Fig. 56.

Common Prawn.

a, Large antennæ; a^2, antennules or small antennæ; b, front leg, with small claw bearing a brush; c, carapace covering the head and shoulders; cl, second leg with holding claw; ab, the ringed abdomen; S, swimmerets; t, tail.

bearing three branches, move gently to and fro in the water. Why do they do this? Because in their last joint where they touch the head, is a little bag beset with hair, and having in it a thick fluid and some tiny particles of sand, and this is the ear of the prawn from which a nerve passes to the main nerve-mass in his head; so that as he moves the antennules in the water, he is, as it were, listening without ceasing to all sounds that may pass through it. Just above these hearing organs a pair of gleaming eyes stand out upon short stalks, and if you examine these under the microscope, you will see that they are composed of a number of six-sided facets arranged in a hemisphere, so that the prawn can keep a sharp look-out on all sides. Here, then, we have an animal with a keen power of sight, of hearing, and of feeling; and if you have ever watched a prawn hunting over the scent of a piece of meat which has been dropped into an aquarium, you will not doubt that he has also the sense of smell, though it is difficult to point out exactly where the smelling organ is.*

And now suppose that he has scented or caught sight of his prey, whether it be a piece of dead flesh or a soft tender living shrimp, he darts down upon it, and seizing it with his second pair of feet (*cl*, Fig. 56), which have large pincers, picks it to pieces with his mouth and claws, and eats it, much as a child eats a biscuit held in its hand, but not with the same kind of mouth. If you will get hold of a prawn and try to make out its jaws, you will

* There has been much discussion as to the position both of the smelling and hearing organs. It seems, however, from Mr. Spence Bate's experiments, that the ear must be at the base of the smaller antennæ, and probably the organ of smell is at the base of the large ones.

be terribly puzzled with the number of pieces in them, for you will find no less than six pairs. The outer pair are evidently altered feet, which are folded right over the others so as to cover them in safely, much as you might put your hands before your mouth; under these lie two more pairs, with little feelers attached to them; under these again are two other pairs, rather differently shaped; and lastly under these a stout pair of jaws, with sharp edges for biting, and a surface for grinding the food. These jaws do not work up and down as ours do, but from side to side like the jaws of a bee or ant, and they are most useful to the prawn in tearing its food.

But how can he have come by so many? Let us look back for a minute to the worm, which you will remember had no true head, but only a long upper lip, and a line of rings on its body, each bearing its own pair of bristles. Now, the prawn also is a ringed animal, only that in his head the separation between the rings is lost, and in his thorax they have grown closely together so that we can only count them by the lines *under* his body, and by the limbs, which grow one pair to each ring. Thus, wherever there has been a ring, there a pair of jointed limbs remains, altered to suit the wants of the animal, and as the head is made up of many rings, these come close together, and form the eye-stalks, the antennæ, antennules, and the mouth-pieces; while the five rings of the thorax bear the five pair of jointed legs, and the swimmerets and tail-pieces spring from the rings of the abdomen.

While all the crustacea keep to this rule of a pair of jointed limbs to each ring, the changes are endless by which these rings and these limbs have

THE MAILED WARRIORS OF THE SEA. 163

been modified to suit their lives. Thus for example, while the prawn uses his second pair of feet for catching and holding his prey, it is the front feet of the crab and lobster which carry the large strong claws, and in the shrimp these front feet have a kind of broad hand at the end, with a hook attached. Again, the skeleton of the prawn remains clear and transparent, but the warlike crab and lobster secrete layers of lime in their skeleton, forming a stony coat.

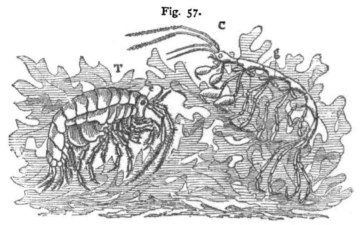

Fig. 57.

T, Sandhopper.* *e*, Flat eye. C, Skeleton Shrimp.† *g*, Breathing gills.

Then again if you look at the nimble sandhopper (T, Fig. 57), with eyes flat in its head instead of being raised on stalks, you will notice that all its body is ringed right up to its head, so that it can bend itself almost into a circle, and flinging back its tail with a jerk, spring about in the sand.

In the skeleton shrimp (C, Fig. 57), which crawls about among the weeds under water, the body has

* Talitrus. † Caprella.

become so thin that it looks like a mere chain of bony rings with legs hanging on to them. In this curious shrimp I want you particularly to notice the little bag-like flaps, *g*, hanging down where the legs join the body. These are its breathing gills, in which the colourless blood of the veins takes up oxygen as they lie bathed in the water. Now, when you next eat a prawn or shrimp, lift up the shield or carapace (*c*, Fig. 56) covering the thorax, and you will find a row of curious bodies (*b*, Fig. 58), looking something like curled feathers, lying against its sides, and fastened

Fig. 58.

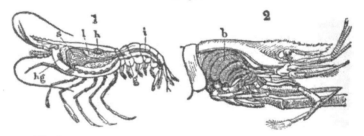

1, Ideal section of prawn, showing, *s*, stomach, below this the mouth; *l*, liver; *i*, intestine; *h*, heart; *g*, chain of ganglia or nerve-masses; *hg*, head ganglia.

2, Prawn with carapace removed, showing gills or branchiæ, *b*.

to the legs. These are the breathing gills of the prawn, and they will remind you of the "ladies' fingers" which we clear away in a lobster before eating it. Though both in the lobster and prawn the shield has grown over and covered these gills, yet you will see that they are really on the outside of the body, at the top of the legs, as in the skeleton shrimp, and that water can easily get to them under the shield. In the oyster, you will remember that hairs or cilia swept the water over the gills (see p. 109); but here, by a most beautiful arrangement, too complicated to explain, the movement of the feet near the mouth empties the water out, and so draws in fresh constantly from the back. And here again notice that animals without back-bones do not breathe through their mouths, but through their sides.

Meanwhile our prawn has been swimming and feeding, and you will scarcely wonder at his activity or his quick senses, when you learn that the same double chain of nerves which we saw in the leech runs also under his body (g, Fig. 58), only that whenever two rings are quite lost in each other, two nerve masses or telegraph stations are also joined into one, so that in the head, for example, a large number have come together, and make powerful head-stations (hg, Fig. 58) of nervous power. His muscles too are firm and strong, and fill nearly the whole of his ringed abdomen, and of his legs and claws, so that though he looks so transparent and fairy-like, he is stronger than he appears.

But now there comes a time when he grows restless and uneasy, and ceases to care for food as he wanders about the rocks on the tips of his toes, seeming rather to be seeking some particular spot. The fact is that it is nearly a fortnight since he has changed his armour, and as he is young and growing fast, it begins to be very tight for him. At last he finds a spot to his liking, and taking hold firmly by his feet, he begins to sway to and fro so as to loosen his body inside its covering. Then all at once a slit opens between his shield and the skin of his abdomen, and gradually his shoulders and head back out, bringing with them antennæ and eye-stalks, legs and feet, as perfect as before, and having even their tiny spines and hairs upon them; then with a sudden jerk he pulls out his abdomen and leaves his clear transparent shell so perfect with the coverings of the eyes, antennæ, legs, hairs, and spines, nay, even with the lining of his stomach and digestive tube, that you might believe the real prawn still stood upon the rock. But no! the creature himself is rolling helplessly over, his soft body being scarcely able to keep itself in position, and if any animal were to seize him now his death-hour would have arrived. He knows this well and soon begins to strike out his abdomen and work his swimmerets which are gradually stiffening and strengthening, and so manages to swim or creep into some sheltered nook, where his inner coat, which has long been forming, hardens, and he is a valiant prawn again.

He is now quite clean and bright and beautiful, and he loves to remain so, and is most particular about his toilet, in fact the prawn is one of the few crustacea which has been seen to brush himself up with great care, though probably many others do it. We have noticed that his strongest claws are not on the front pair of feet as in the crab and lobster, but on the second pair. The front claws are fine and delicate, and carry little brushes on their tips; and the prawn has been seen standing on his four hinder pair of legs with his tail tucked under him, and using his front pair to brush his swimmerets, afterwards passing them through his foot-jaws to clear the dirt off the brushes!*

---

* The little broad-claw crab cleans himself with the hind pair of feet instead of the front ones.

THE MAILED WARRIORS OF THE SEA.

Not so the large crabs, the backs of which we so often find covered with weed and shells and small tube-worms which have settled upon them, so that when a crab is old and does not change his shell, he often carries a perfect colony of life about with him. If the prawn is the crystal fairy of the sea,

Fig. 59.

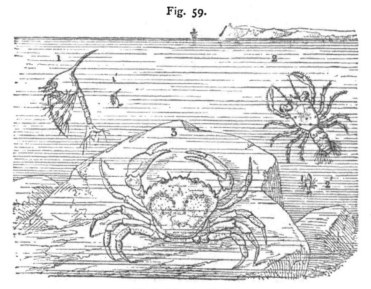

Early life of a Crab.

1, A crab soon after birth; 1′, natural size. 2, A crab after changing its skin several times; 2′, natural size. 3, The young perfect crab after it has tucked its tail under the carapace.

surely the crab, when big, is the lumbering armed giant, who destroys and devours without mercy, glaring out of his coat of mail, and not fearing any creature except a stronger crab than himself. He spares no animal, whether fish, mollusc, crustacean, or worm, that comes in his way as he sidles along on his strong muscular limbs; but seizing them in his powerful claws he cracks their shells if they have any, and tears their flesh to pieces, tucking it greedily into his mouth, which looks as if it were in the middle of his body. He even makes no difficulty of breaking the shell of one of his own kind and feeding upon it from behind, while it in its turn is eating some smaller and weaker brother.

To devour and be devoured seems to be the main mission of crabs, and they feed so greedily that we shall not be surprised to learn that besides their array of outer jaws, they, and many of the other crustacea, have hard teeth in their stomach (s, Fig. 58) which help to grind down the food. You may see these teeth well in the stomach of the lobster, where children often call them the "lady in her chair."

At first sight it is difficult to understand how a crab can belong to the ringed animals, but if you lift up the tail, which is tucked under the body, you will see that this is ringed like the abdomen of the prawn, and if you break off the legs carefully you will find under them the finger-like gills, showing that the body of the crab answers to the head and thorax of the prawn, only that the shield over its back is much broader, and is fastened down firmly at the sides, while the tail is tucked under instead of standing out.

Moreover, if you could see the crab when he is first hatched from the egg (1, Fig. 59) you would see his tail stretched out and jointed as distinctly as that of the prawn, and at this time, with his flat eyes and a curious spine sticking out of his back, he is as unlike a crab as can well be. In this state he swims about vigorously, and in seven or eight days, having cast off his coat several times, he loses his spine, his back becomes broader, and he becomes a tailed crab (2). Still he goes on swimming and clinging to seaweed or anything he can find, till, after moulting a few more times, his tail is folded under and he sinks to the bottom a true walking crab (3).

A change or *metamorphosis* of this kind takes place in nearly all the crustacea during their growth, though it is different in the various forms.

After the crab has assumed his real shape he lives on the bottom of the sea, generally in deep water, and in the holes of the rocks, and fights bravely for his life among his companions. Only about four times in the year while he is young a season of fear and anxiety comes upon him, for his shell will not allow him to grow any larger and he must part with his strong armour. Then he creeps into the darkest hole he can find, and, throwing himself upon his back, swells out his body till he forces his covering shield to break away from the under part, and so he creeps out. He does this with much pain and difficulty, for his claws are much larger than the joints through which they have to be pulled, and they are often cut and lacerated in the process. He could not, in fact, get out at all if it were not that his flesh becomes watery before he casts his shell. Every housekeeper knows and avoids buying a watery crab, in which the flesh is poor and thin and the shell is half filled with fluid. When his shell is cast the crab waits trembling in his hole for a new layer of lime to form before he can venture boldly out again. It is said that at these times when a mother crab loses her shell and becomes soft, her mate will watch the hole where she is lying and keep her safe till her shell has hardened.

But how, then, is it with the hermit-crab? He, poor fellow, never loses the long tail which all young

Fig. 60.

Hermit-crabs.*

1, The hermit-crab in a whelk-shell walking. *c,* The large claw which closes the hole when it retreats into the shell; *f,* smaller feet.

2, The hermit-crab coming out of the shell. *a,* The soft abdomen; *h,* hooks by which it takes firm hold in the shell.

crabs have when they are born, and, moreover, the skin which covers his abdomen is quite soft, thus always offering a tempting morsel to hungry sea-animals. One would think that here was a disadvantage very unfair to the half-naked animal. But wait a mo-

* *Pagurus Bernhardi.*

ment and consider how many thousands of hermit-crabs of all sizes feed on the dead fish and garbage of every sea-shore ; and how well they are protected by the strong winkle and whelk shells which they choose for their houses, so that they can hold their own, when the tiny crabs wearing only their own brittle coats, would soon be cracked and eaten. Evidently the hermit-crab has found stolen houses an advantage to him, and the way in which his tail has become adapted to his home, while keeping all the usual parts of a crab, is most curious.

One of his claws (c, 1) is much bigger than the other, and closes the opening of the shell after the rest of the body is drawn in, thus barring the door against most intruders, although the fiddler-crab sometimes manages to thrust in his thin pincers and pinch the hermit to death. His next two feet are strong, though pointed, and are able to take a firm hold on the sand as he walks and to bear the weight of the shell, while the two comparatively thin pairs which follow serve to shift his body in its house. His swimmerets, no longer needed, are stunted and small, and his soft abdomen follows the winding of the shell in which he lives ; while the tail fin, no longer broad and flat, is turned into a kind of grappling hook (h), which takes hold so firmly that he is scarcely ever dragged out alive. So there is but little danger for him except when he is changing his shell for a larger one, and this he does wonderfully quickly, never leaving his old house till he has found a new one. In fact the hermit succeeds so well in life that he is extremely pugnacious, and will soon make great havoc in an aquarium. Moreover, he often feeds two other animals besides himself, for a parasitic anemone often lives upon his shell, and a beautiful worm* shelters within it and has been seen to tear the food out of the crab's mouth.

So each in their several ways, the prawns, lobsters, and crabs, struggle for their livelihood. Brave, hardy, and voracious, they spare scarcely any creature of the sea of moderate size, whether dead or living, and they fight so madly, that fishermen sending lobsters alive to London, are obliged to run a piece of wood in the joints of the claws to prevent them from maiming each other on the road.

They care but little for lost limbs, for these will grow again; and when wounded, so that they might bleed to death, they throw off the shattered limb at the next joint, where a new skin quickly forms, and the danger is averted. No doubt hundreds die both in youth and age, yet the multitudes never diminish, for one lobster alone will produce 20,000 eggs, which she will carry patiently for six months under her abdomen, fastened together by gluey threads. Even after she has broken open the eggs by the movement of her tail, and released the baby lobsters, she will still carry them till their coat is hard and firm, and only then leave them to wander alone. The crab and the prawn, on the contrary, turn their little ones out at once to swim as scarcely visible specks in the open sea, where they feed and grow till their strange changes of shape are worked out.

The crab family, however, are not satisfied with one kind of life; the velvet fiddler-crab of our shores

* *Nereis bilineata.*

has its hind feet broad and flat, and may be seen swimming, when the common crab can only creep; while the oceanic crab has taken to the open sea, and can swim for days without resting, feeding the while. Then, on the other hand, there are crabs living on the land. The racing crab of Ceylon, which will outstrip the swiftest runner, burrows in the dry sand; and though it likes to have its gills moist, dies if held under water. The frog-crab of the Indian ocean climbs on the roofs of houses; the robber crab of the Mauritius lives in holes lined with cocoa-nut fibre at the roots of the cocoa-nut palm, and breaking open the nuts feeds upon the fruit; while the land-crab of the West Indies burrows in the ground, and goes only once a year to the sea to lay her eggs. Still all these crabs retain enough of their old habits to like to have their breathing-gills wet, and most of them visit water daily for this purpose, while some of them have a curious way of keeping the water enclosed, and freshening it with air, while others use the water till it is exhausted, and then raise their shield or carapace and breathe as land animals.

And now after hearing of these land crabs, we shall not find it so difficult to believe that the little wood-louse of our gardens, which curls itself up like a ball, and is the only form we have remaining like the huge trilobites of ages gone by, is a true crustacean, adapted for breathing air though still loving moist places.

But we must return to the sea, where a most curious and interesting group still remains for us to study. We have heard of old families among men, who, having met with misfortune, have had the good sense to set to work and earn their daily bread in quiet obscurity; and among the lower animals we have seen that many, like the sponge and oyster, give up the free roving life of their childhood, and settle down upon one spot. But who ever before heard of a creature, which, after swimming about in a rational manner with an eye or eyes to see with, and antennæ to feel with, behaving like an ordinary and respectable individual, should put its forehead down to a rock and cement it there by means of glue from its antennæ, and should remain thus all the rest of its life with its head downwards and its heels in the air, kicking its food into its mouth.‡

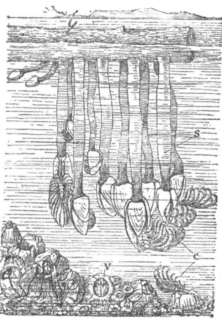

A Group of Floating Barnacles * with a bank of fixed Acorn-Barnacles † in the foreground. *s*, Fleshy stalks growing from the head of the barnacle; *c*, cirrhi by which the animals feed; *v*, the inner valves of the acorn barnacle which open and close.

* Lepas. † Balanus.
‡ Huxley, *Anatomy of Invertebrates*, p. 294.

Yet this is the true history of the barnacles and acorn barnacles of our coasts, and nothing can explain such extraordinary behaviour, except the overcrowding of the sea, and the struggle for life, which drove these curious creatures to prefer feeding upside down in places where others left room for them, to starving in an upright position.

It is worth while to spend a short time sitting by a seaside pool or on the steps of a pier to watch the animals within the white shelly cones, called acorn-shells (Fig. 61, foreground), feeding in the sea-water. Each little cone is made of a number of shelly pieces, and in the middle of these you will see from time to time two valves (v) open, through which a tuft of feathery transparent fingers (c) is thrust out, looking like a curl of delicate hair.* Then after opening out in the water, the curl is drawn up again just as you are beginning to admire it, and the valves close. Not for long, however, for almost immediately they open again and the same process is repeated; so that in a group of acorn barnacles all is in motion as one after another sweeps the sea for food.

These tufts are in fact the fringed legs of the *balanus*, as the creature is called; and looking at him as he is fastened down inside the shell, you will see that he is something like a rough attempt at a shrimp, lying on its back, mouth uppermost, so as to be able to seize and devour the minute creatures of the sea drawn in by the fringes of the legs.

In this way safely ensconced in his jointed shelly carapace of carbonate of lime, it is easy to see that

* These fingers are called *cirrhi*, from *cirrhus*, a curl or lock of hair.

the balanus lives more securely than if he had remained a freely roving creature as we see him at 1, Fig. 62; and the success of his retirement to a fixed life is proved by the countless number of acorn-shells which are found on every sea-shore. The *lepas* or barnacle with stalks (Fig. 61), you will see less often, for they live in deeper water attached to rocks or pieces of floating timber. Their history is the same as that of the acorn-shell, only that from the cement of their antennæ they form long fleshy stalks which fasten the head to its support.

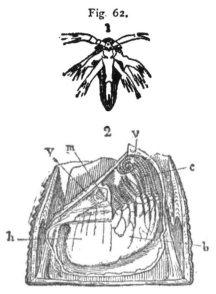

Fig. 62.

Development of an Acorn Barnacle.

1, A young acorn barnacle before it has settled down.—*Spence Bate*.

2, The grown-up acorn barnacle cut in half; *h*, head cemented to the shell; *b*, back; *m*, position of the mouth; *c*, cirrhi fringing the legs which sweep in the food; *v*, valves of the shell.—*Darwin*.

And here we must take leave of the crustacea. We have really only made acquaintance with three branches; 1st, the ten-footed and stalk-eyed crustaceans, the prawns, and crabs; 2d, those with eyes fixed in the shell and breathing-gills fixed to the legs, the sandhoppers, and skeleton shrimps; and 3d, the barnacles. But these by no means represent even the chief forms. The King-crab* of the Moluccas, with his horseshoe carapace and spiked tail, represents a whole race which flourished long before the coal-forests grew. The beautiful little fairy-shrimps† of our ponds are another type whose feet are used as breathing-gills, while the tiny water-flea‡ and cypris of our ponds are true crustaceans, though they have two-valved shells like a scallop.

The histories of these little beings are as yet not much known, and in truth it is impossible to follow out all the strange vagaries of the crustacea without making them a life study. Even when we have exhausted those which live independent lives, there remains a whole mass of parasites which fix themselves on the backs and in the gills of fishes, and even under the tails of their distant relations the crabs and lobsters.

Thus we find that these "insects of the sea" have spread everywhere where there is water, and have even found their way on to the land. Yet they are scarcely likely ever to make much way on dry ground, for we have seen that they have always a lingering tendency to breathe water, and therefore they are at a disadvantage among the myriads of insects fitted for the air. Meanwhile, though we may know more of the habits of the spider and ant, than of the crabs and the barnacles which hide from us in the ocean, yet those who love to study complicated family history will find no class in the animal kingdom with such an interesting and involved genealogy as that of the *Crustacea* or crust-covered animals.

* Limulus. † Branchipus. ‡ Daphne.

CHAPTER IX.

THE SNARE-WEAVERS AND THEIR HUNTING RELATIONS.

> So dangles o'er the brook, depending low,
> The spider artist, till propitious breeze
> Buoy her athwart the stream. From shore to shore
> She fastens then her horizontal thread,
> Sufficient bridge, and traversing alert
> Her fine-spun radii flings from side to side,
> Shapes her concentric circles without art,
> And, all accomplished, couches in the midst,
> Herself the centre of her flimsy toils.
>
> HURDIS.

IT was a hot spring night on the coast of the Mediterranean in the south of France, and the hum of the night insects filled the air. The night-beetles were flying hither and thither, and the crickets on the terraces of the olive-groves were loudly chirping their love-songs. One in particular, whose dark brown body could scarcely be distinguished against the bank even by nocturnal enemies, was working his wing-cases with a will and sending out a clear and piercing cry. He little thought that he was sounding his own death-note, but so it was, for behind him from under a large stone in a damp corner in the side of the bank, an enemy was stealthily approaching.

Any one who had been lately studying prawns and lobsters in the sea on this same shore, would almost have fancied that this enemy was a curious small lobster which had come upon the land, for two large claws were stretched on each side of his head, and with them he felt his way as he crawled along; his

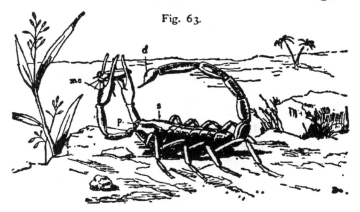

Fig. 63.

Scorpion with a Cricket in its claws.
s, Shield or carapace; *m c*, mouth claws; *p*, pincers; *d*, poison dart.

jointed feet were cased in horny armour, and so was his whole body, which had a shield over the head and shoulders as a prawn has; while his tail, which dragged heavily behind him, was covered with the same kind of horny rings. Two large eyes, with some smaller ones near them, shone in the front of his head, and he was slowly but surely advancing upon the unconscious cricket. And now, he was close upon him, and in a second, almost too quickly for the movement to be seen, the long claws were thrown forward, and the cricket was seized and held up in the air. But our friend had no intention of yielding so easily, he was strong, and he struggled violently—in vain, for his captor briskly curled up the long tail which had till now seemed such a burden to him, and from the tip of it thrust a poisoned dart into the body of his victim. So the sturdy cricket died in the grasp of the Scorpion.

Nor was his captor long in devouring his prey. Bringing the cricket down to his mouth he pierced his skin with the sharp pincers (p, Fig. 63), which take the place of antennæ on his head, and soon sucked out the juices of his body; then dropping the empty skin he went dragging slowly on his way, in search either of fresh food or to find some mate wandering like himself.

Plenty of these fierce little Scorpions, which hide under stones by day and come out by night, may be found in the warm sunny south, and though they look so like crustaceans, they are true land animals. They have no means of spinning, and have a poison dart in the tail quite peculiar to themselves, yet they belong to the spider family, as may be seen by their eight pairs of legs, their sharp pincers which take the place of the antennæ of insects, their claws which are part of their mouth-pieces and are fixed to the jaws, and the narrow slits under the abdomen through which they take in air to breathe. They lead but a lonely life; for whether in the sandy plains of Africa, where they are often as much as a foot long, or in the burning heat of South America, or on the warm bright shores of Italy, each scorpion burrows under his own protecting stone, rarely having any other with him. Even in summer it is only at night that they seek companionship; while in the winter they burrow deep in the ground and sleep till the warmth comes round again.

While the spider is the industrious and skilful snare-weaver of its class, the scorpion is the fierce bandit, knowing well the power of its sting and the terror it inspires; and like the bandit it lives in a state of perpetual warfare, flinging its tail over its head and extending its claws at the least alarm, and either fighting till death or running rapidly backwards facing its foe, till it reaches a place of safety. And in like manner as the robber's wife, shut out from the companionship of the rest of womankind, will love and defend her children with wild devotion, so the female scorpion will carry her young brood for many weeks after they are born, clustering all over her back, till they are able to fight for themselves.

We must not, however, pause long over these solitary and dangerous creatures, for a far more interesting group of the spider class flourishes here in our own country, where all who wish may study its members.

Instead, therefore, of lingering in the warm south, let us return to England, where, in the cracks of some old paling, or under the leaves of a shrub on a summer's evening after some days of thunder and rain, we may find a common garden spider * lying crumpled up as if half dead. Her web, long ago destroyed by the wind and the rain, has left her no means of getting food for many long hours, and she

* *Epeira diadema.*

is waiting patiently till the weather will allow her to spin fresh snares. How mean and shrivelled and helpless she looks, any one will know who has ever found a spider in this deplorable condition; and certainly no one at first sight would imagine that this crumpled-looking object could have the ingenuity and skill to weave the web which a few hours later will be stretched across the bushes. And yet, as set free from your hand she hurries away, scrambling over the ground in the twinkling of an eye, or dropping nimbly by means of her almost invisible thread, there can be no doubt that she is both active and intelligent; and a little patient examination will show that the poor despised spider, which for some unknown reason is so often disliked by mankind, is one of the most industrious and cleanly, skilful and patient of life's children; while she carries upon her body some of the most curious implements ever devised, for doing her work in the world.

Look at her limbs in their jointed casings (Fig. 64), and you will see that here is the same outside horny skeleton as in the prawn and the scorpion, with elastic skin between the joints; but her abdomen (a) has almost entirely lost the traces of rings, and is often covered with fine down; while her head and shoulders, welded like theirs into one piece (t), are sturdy and strong, giving her great advantage in attacking and devouring the numerous insects which fall victims to her bloodthirsty appetite.

Perhaps you will think at first that she has antennæ, for two short feelers ($p\,p$, Fig. 64) stand out in front of her head. But these, like the claws of the scorpion, are part of her jaws (j), and are fastened on to them. None of the spider family have any true antennæ. In the scorpion we found them turned into pincers (*p*, Fig. 63); in the spider they

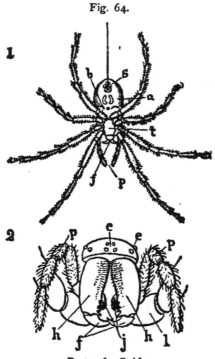

Fig. 64.

Parts of a Spider.

1, Under part of a spider's body. *t*, Thorax or chest from which the eight legs spring, and to which the head is united in one piece; *f*, fangs; *p*, palpi or feelers attached to the jaws; *a*, abdomen; *b*, breathing slits; *s*, six spinnerets with thread coming from them.

2, Front of spider's head. *e*, Eyes; *p*, palpi; *l*, front legs; *h*, hasp of fangs; *f*, poison fangs; *j*, outer jaws.

have become most dangerous and powerful fangs (*f* 1, and *h, f* 2, Fig. 64), which hang down over her mouth; and while the scorpion carries her poison in her tail, the spider bears it in her head, and, as we shall presently see, pours it into her victim through these pointed weapons (f).

But now, while we are talking of her, our friend the spider is beginning to grow restless, for as evening draws on and the temperature of the air and other signs promise a fine night, she is anxious to spin a new web, so as not to go supperless to bed. Have you any idea how she does this? for she has nothing but her own body both to supply the material and the machinery for the work.

Look carefully under her abdomen, and near the tip you will find six little nipples (s), looking something like miniature copies of the teats of a cow. Under these nipples, inside her body, there are special glands, in which a kind of gum is secreted, and this dries when it comes into the air, and forms the silken thread from which the spider hangs, and out of which she forms her web. And now comes the almost incredible part of the story. These nipples, which are called "*spinnerets*," have not merely one opening like a cow's teat, but each one, tiny as it is, is pierced with at least a hundred holes, and when the spider begins her web, *more than six hundred separate strands* go to make the one slender thread which you see stretched out from her body. The four spinnerets nearest the tail give out the long threads, the two above, moving from side to side, weave the whole into one connected line. Nor is this all, for the spider can close any of the holes at will, and a fine or a coarse, a dry or a spangled thread comes from her body, according to the use she wishes to make of it.

But when the thread is made, how is it to be drawn out and guided on its way? Under the microscope her feet are seen to be formed of three claws, the middle one longer and bent, so as to grasp the threads as she runs, and the other two toothed liked combs. With these combs and the spines and hairs upon her legs she manipulates the tender

Fig. 65.

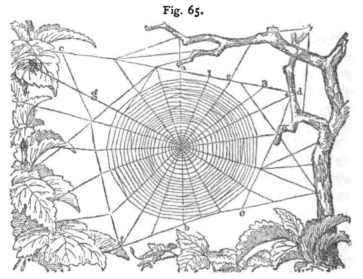

Web of the Garden Spider.

thread as it oozes from her body and does with it what she will.

And now let us watch her at her work. Running hither and thither, she first selects a spot (*c*) to start from, and pressing her spinnerets against it leaves there a little lump of sticky silk. Then standing quite still she gives out from her spinnerets a light floating thread. Longer and longer it grows and floats in the evening air, and by and by the end catches on some bush, let us say at the point *d*. Instantly the spider feels the pull, and fastening the end out of her own body to the sticky lump *c*, she runs along the line *c d*, strengthening it with silk as she goes. That line being firm she can use it as her tight-rope, and running from bush to bush may either let fresh threads float, or carry them with her from point to point till she has made a square or triangle of threads (*c, d, e, f*), anchored here and there to the leaves and stems. And now she must make the cross lines, so going to the middle (*a*) of the highest line she drops down, and swinging to and fro catches the lower line and stretches a vertical cord (*a b*).

Then she is ready to begin the web. Up to this time she has been obliged to work slowly and with many difficulties, but from this moment the work goes on apace. Running to the middle of the line *a b* she fixes her thread there and then goes on to the other side, carrying the line with her and keeping it carefully with her feet from clinging to the one on which she runs. Arrived at the stretched line she fastens this loose thread to it at 1, and so makes one of the spokes of the wheel. Then moving a little farther along she fixes another end at 2, and running back to the centre forms another spoke; and so on through 3, 4, etc., till all the spokes are made. Then she goes back to the middle, and walking carefully round and round the spokes lays down a winding thread from the centre to the outside, fixing it to each of the spokes by a minute drop of gum. At first this thread is dry and hard, but when the spider is at a short distance from the centre she changes her material and gives out a beautiful fine thread, spangled at every point with minute drops of gum, which will not harden in the air;* so that by the time she has reached the end of the spokes she has left behind her a glorious spangled web, closely woven and wonderfully elastic, because the drops of gum yield gently as the web sways, while they are so sticky that no insect flying against the web is likely to get away again.

In this manner, in about three-quarters of an hour, the nimble little spider has woven a snare measuring perhaps half a yard across and spangled with more than a hundred thousand gum-drops, and there is yet time for some of the late flies to be caught before night comes on. She has left the middle of the web with its dry scaffolding thread, because it is there that she hangs head downwards waiting for her prey. Sometimes, however, she will prefer hiding herself under some leaves in a bush, and then she will carry with her a strong thread (*g*) attached to the middle of the web to give her warning of any disturbance. And now a good-sized fly comes buzzing along, and running its body against the web gives it a shake. Instantly

> "The spider's touch, how exquisitely fine,
> Feels at each thread and lives along the line,"

and almost before you can see her, she has darted

* Mr. Emerton, an American naturalist, who has watched the Epeira at work there, states that she lays down first a dry scaffolding from the centre to the circumference, and then working back again, destroys this hard thread as she lays down the spangled one. Mr. A. Butler and Mr. Lowne, however, both assure me that they have often seen the English garden-spider spin her web, and that she invariably lays down the gummy thread at once.

from her hiding-place to the centre of the web. Here she herself gives it a shake to find whether it will be answered, showing that a live object is causing the disturbance. The unfortunate fly quivers in the toils, betraying its whereabouts, and straight the spider darts upon it, and with one sharp bite ends its life. It is not however, strictly speaking, with her mouth that she has bitten it, but with those two poison fangs which we spoke of just now, which hang down over the mouth. While she was spinning her web, or patiently waiting under the leaves, these fangs (*f*) were shut into the cases above them (*h h*) just as a clasp knife is shut into its handle; but directly she seized her prey they were opened, and the sharp points driven into the fly's body gave out poison from their tips, and quickly put an end to its life.

And now, being hungry, she seizes the dead fly with the two feelers or palpi of her jaws, and holding it to her mouth sucks out its tender juices; and she has no need to pause for breath, for you will remember that she does not breathe through her mouth. It is under her abdomen that you must look for the two narrow slits (*b*), through which air is taken into sacs within her body.

But another interruption occurs. While she is still busy with her meal, a fresh shake of the web informs her that a new victim is caught, and she hastens to the spot. This time it is a strong night-beetle which is caught in the toils, and she cannot grapple with him so rashly as with the fly, while his struggles threaten to break the net. In this dilemma she has a stratagem ready. Pressing her spinnerets against the web, she begins to weave round him a covering of silk, and going closer and closer as his legs are entangled, she twists him round and round with her feet till, quite enveloped, he can struggle no longer and receives his death-blow.

But if by chance it had been a wasp, and she dreaded its sting, she herself would have torn the strands of the web and let it fly away sooner than run the risk of being the conquered instead of the conqueror.

So the garden spider lives, spreading her snares with wonderful skill and care, running and dropping from her thread with agility and precision, and making great havoc in the insect world with her poisonous fangs; and if you ask what apparatus she has within to guide all these wonderful actions, we must go back again to the knots of nervous matter which we found in the body of the leech. For here in the spider we find that many of these knots are clustered thickly together in the head and neck, forming what might almost be called a brain, and connected with the line of knots running along in the under part of the abdomen. Thus, while the spider is endowed with many tools and weapons upon her body, she has also a strong battery of nerve power within to govern them.

Nor can we doubt that she owes to the stern lessons of life much of her skill and intelligence. She has to undergo sad privations, and in bad weather, when starvation stares her in the face, she is often driven to wander in search of such insects as she can catch; while long experience has taught her race never to spin a web when it would be destroyed by wind or rain, but to fast patiently or trust to seizing any small insects coming within her reach rather than to use the small stock of silk which cannot be replenished till she gets fresh food.

You may perhaps wonder that all this time we have spoken of the spider as "she," but the truth is that the female spiders do most, if not all, of the work, for the male spiders are much smaller, and very rarely spin webs. They seem to live much at the expense of their wives, and, sad to relate, are very often killed by their spouses when these are tired of their company.

And now, if the female spider succeeds in getting a living and escaping the birds and other enemies until the autumn, she spins a strong cocoon of yellow silk which she secures under some stone or into a crack in the wall, and though it measures scarcely half-an-inch across, yet she manages to pack into it from six to eight hundred eggs, and then leaves it; and next spring, when the warm weather comes, the young spiders struggle out of the eggs, and working themselves free from the skin which hampers their limbs, cling together in a ball for about a month, and then separate and begin to spin webs as their mother did before them. They cast their skin many times before they are grown up, and even afterwards they creep out of it once a year and begin again with a fresh bright coat.

From this garden spider we have learnt to know roughly the manner of life of the spinning spiders, and the tools with which they work; but their devices for gaining a living, the nature of their webs, and the different nooks and corners they find for shelter, are almost endless. Look at the common house-spider instead of carelessly sweeping her web away, and you will find that she lays her threads roughly in all directions in the corner of the room, running from one to another till she has filled it up with a fine web, not sticky, but so entangled that the flies catch their feet in its meshes. In one corner you will find a little silken tube like a thimble which she has made as a house to hide in, out of sight of her prey. Her web will last for many weeks, while the garden-spider must spin afresh or mend her web every twenty-four hours, but on the other hand the house-spider is less likely to have an abundant supply of insects and her web is often ruthlessly destroyed. She will sometimes live from six to eight years, and each year she lays her eggs in a cocoon and hides them in a tuft of silk thickened with scraps of whitewash and plaster, and broods over them till the young ones are born. In the walls of some outhouse or warm greenhouse you may often find small spiders' webs in the summer time, with three or four cocoons in them, and numbers of tiny spiders creeping out upon the web.

Then look carefully on a summer's morning among the gorse and heath of a common, and you will find delicate webs spread almost on the surface of the ground. Instead of trampling these under foot, seek out the centre of each web, and there, in many cases, you will find a hole leading straight into a tunnel in the ground. This tunnel will be lined by a tough web, while at the bottom the little spider will be crouching, her feet resting on the threads, and ready in a moment to dart out when the toils are shaken. This spider has learnt how to hide herself from the birds, the squirrels, the frogs, and the toads, which devour her neighbours, while at the same time she spreads her nets and catches the beetles or the

Fig. 66.

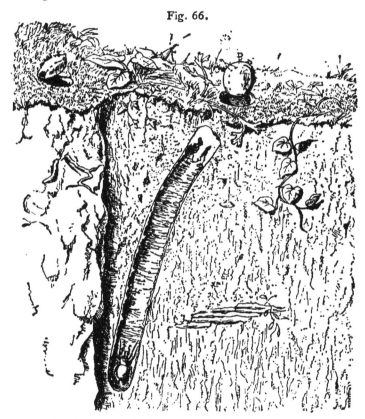

Nest of one of the Trap-door Spiders.—*Moggridge.*

D, The door which closes naturally by its own elasticity and weight; *m*, marks of the spider's claws where she has held it down from inside.

midges for her daily food. If when wandering along the country lanes you look carefully in the loose soil of the bank, or the crevices of old stone walls you can scarcely fail to find some of these tiny webs leading into silk-lined holes where the spider is waiting for her prey.

Yet even these are not the most clever of all spiders, for on the shores of the Mediterranean you may find some which not only live in silk-lined tunnels but actually make doors to their houses (see Fig. 66). These doors are made of layers of web and earth, and they shut down naturally by their own weight, so as to be quite hidden by the grass growing over them; but, if by chance they are disturbed, the spider herself will often rush to the top of the tube and sticking her claws into the door (D *m*, Fig. 66), will hold it down with all her might as she presses her body against the sides of her home.

Now see how this spider gains her living. A naturalist named Erber once sat out for many hours on a moonlight night watching her doings, and soon after nine o'clock he saw two of these spiders come out each from their holes, and pushing open their doors, fasten them back by fine threads to the blades of grass near, and then spin a web round the open hole and go back into their tunnels. By and by two night-beetles were caught, one in each web, and in an instant the spiders darted out and pierced their victims with their poisoned fangs, sucked out their soft flesh and then carried the empty bodies away to some distance from their holes. Then Erber left them, and in the morning the spiders had cleared away all trace of the webs and were shut down snugly in their hidden homes.

Which among us works more cleverly or with

more industry for daily bread than these little spiders? and they do it too among many difficulties and dangers, for the birds and lizards are watching above ground to make a meal of them, and the centipedes and other crawling insects creep into their holes to attack them. Some of them have learned a means of escaping even this danger, for they make a second tunnel branching out of the first, and build a doorway between the two so that they can retreat into the second passage in case of attack, and setting their back against the door baffle the intruder.

So in the air and on and under the earth, the spiders spin their webs, and since they must try every means of gaining a living in this struggling world, there are some, such as the wolf-spiders, which, instead of spinning webs and waiting for their prey to come to them, search for it among the low bushes and leaves and grass, and use their spinners chiefly for letting themselves drop from a height, or for spinning their cocoons, and lining the holes in the walls where they retire for their winter sleep. You may find these running about in the woods and on heaths, and if you catch them about June you will find each one carrying a snowy-white ball under her body. This is her cocoon, containing about a hundred eggs, and if you try to take it away she will fight for it as courageously as any human mother. I took away one three times from a mother on Keston Common last summer, and each time she seized it again, and went off gaily with it at last, none the worse for the struggle.

Then there are the leaping-spiders, which pounce

THE SNARE-WEAVERS.

upon their prey, creeping slowly along a wall and sliding nearer and nearer, till suddenly they leap, and seizing the victim kill it, and return by the silken thread which connects them with the wall above, and saves them if they fall too far. These spiders often roll up their cocoons in the leaves of some bush where you may find them in the early spring.

The hunters and leapers can often find food which does not come in the way of the web-spinners, and when all the domains of earth and air are overrun, then there are other kinds which take to the water. How few people think as they walk through quiet country-lanes, that in the deep watery ditches often to be found near rivers which run in low ground, a little water-spider may be living, coming to the top to breathe as a diver does, and carrying down air-bubbles entangled in the fur which covers her body and between her legs, and so filling a curious domed hall which she has built in the water below.

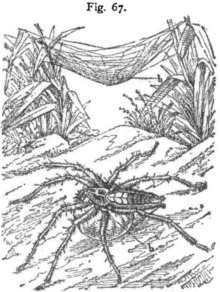

Fig. 67.

A Hunting Spider* with a bag of eggs, *b*.
Staveley.

* *Dolomedes mirabilis.*

It is worth some trouble to find this wonderful little creature, which has often been seen in the fens of Cambridgeshire, and in the ditches near Oxford, and also in Ireland, and which, though fitted to breathe in the air, has learnt to take refuge in the water, and find there her food and her home. She fixes her house on the stem of some water-plant, spinning there a thimble of delicate silk into which she carries air, shaken off as bubbles from her body; and this air rising up to the top of the thimble gradually displaces the water and fills the whole chamber. And so in peaceful but not entirely stagnant water,

> On light sprays hung,
> By silk cords slung,
> O'er-arched by a silken dome,
> Is the airy hall
> With waterproof wall
> Where the spider makes her home,

and there she lives quite dry, and spins her silken cocoon with its hundred eggs, out of which come the young spiders which begin at once to build and live as she does. Even when she makes her journeys to the surface to catch water-flies and other insects, or to take breath, the water does not wet her, for the bubbles of air which glisten over her body, making it shine like quicksilver, keep her skin dry.

And here we must take leave of the true spiders, which roam all over the world, and range in size from the huge hunting spiders of South America and Ceylon, whose legs will cover a foot of ground, and who have been seen to prey upon young birds and lizards, to the tiny red money-spinner, which is so

THE SNARE-WEAVERS.

light that, like others as small as itself, it is often carried up in the air as its thread is caught in the light breeze. It is probably from the threads of these tiny spiders that the gossamer webs are formed, which may be sometimes seen on a bright summer morning hanging in the air entangled in each other, either empty or with their owners within them.

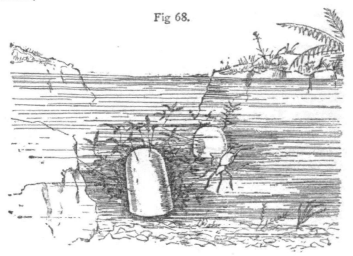

Fig 68.

Water-Spider * with its dome-shaped cell full of air.

Whether they are large or small, however, we find in all spiders the same poison fangs, the same complex thread, and the same scrupulous neatness and cleanliness, which makes them keep every joint and hair of the body free from dust and dirt, and leads them to avoid any dead and decaying food with disgust; and a quickness of intelligence which

* *Argyroneta aquatica.*

the mites is full of interest, and if we could watch them all in their daily labour we should find them probably quite as active and industrious as man himself, and in many cases struggling far more bravely and patiently against the thousand dangers and privations which threaten them at every moment of their lives.

CHAPTER X.

INSECT SUCKERS AND BITERS WHICH CHANGE THEIR COATS BUT NOT THEIR BODIES.

> Yet hark ! how through the peopled air
> The busy murmur glows,
> The insect youth are on the wing,
> Eager to taste the honeyed spring
> And float amid the liquid noon.
> Some lightly o'er the current skim
> Some show their gaily gilded trim
> Quick glancing to the sun.
>   GRAY.

IF any of us were asked the question "What is the use of plants?" I think there is little doubt that we should answer "To make the world beautiful, the air pure, and to provide food for man and beast." But if the same question could be asked of the little green aphis clinging on to the stem of a rose-tree (see Fig. 69), he would know nothing of the beauty or the pure air, nor would he think of man or beast, but he would probably answer that " plants are made for plant-lice." And from his point of view he would be right, for there is probably not a single herb, or shrub, or tree in the world which has not its own peculiar insect, sucking the sap and living upon the sweet juices. We cannot do better than begin our history of the six-legged insects by these little plant-suckers which we all know so well, calling them "blight insects" when we find them covering our rose-trees, or our geraniums in the greenhouses, or our apple-trees in the garden.

Fig. 69.

The Rose-Aphides, or plant-lice.

A, The wingless aphis, like those on the bud but enlarged. *t*, Honey tubes. B, The winged aphis, enlarged. B', Same, natural size. *g*, Blind grub feeding on the aphides.

You may easily find them huddled together on a stem or bud, raised on their six slim legs, with their heads close down to the plant, and looking sleepily out of their two little brown eyes Yet they are by no means asleep but very busy, for their mouth, which, like that of all other insects, is composed of six parts, is so formed that they can plunge it deep into the stem and suck and suck all day, filling their round green bodies with sweet sap. A wonderful little mouth it is, the two lips being joined together into a kind of split tube, out of which are thrust the four jaws, in the shape of long thin lancets, to pierce the plant. When once these insects have fixed themselves they never seem to tire of sucking, but take in so much juice that, after passing through the body, it oozes out again at the tail and the tips of two curious little tubes (*t*, Fig. 69) standing up on their backs. This juice, falling on the stems and leaves of the plant, covers them with those sticky drops often called "honey-dew."

It cannot be said that these little insects lead very exciting lives, for they make no homes, neither do they take any care of their young ones, and only move about when they wish to fix upon some new spot; and yet they are very interesting, both because the ants visit them to sip their sweets, as we shall see in Chapter XII., and also because they manage to live in such numbers in spite of being so helpless and stupid.

The secret of this is that they have a special way of sending young ones into the world. If you look at a stem covered with plant-lice towards the end of the summer, you will find among the wingless sucking insects some larger ones straggling about the plant (B, Fig. 69) which have delicate transparent wings. These are the fathers and mothers, whose wings have grown gradually under their splitting skins, and they will lay the eggs to be hatched next year. But if you look in the early spring you will find no *winged* plant-lice or aphides, but only the little round-bodied green forms, and yet new ones are constantly appearing! Where do these come from?

Do you remember how the hydra of the jelly-fish went on budding and budding for years before another true egg-bearing jelly-fish appeared? (see p. 64). Now we have an insect doing the same thing, for as soon as one of these wingless aphides is about ten or twelve days old, there come from her body not eggs, but young living aphides like herself, three, four, or even seven a day, which struggle over the backs of their companions till they find a clear spot on the stem, where they fix their mouths and suck like the others, only moving to struggle out of their skins about three or four times a day, till they are full grown. Then these young ones begin to bud in the same way in their turn, so that in a very short time from two or three mothers a whole plant is covered. In fact it has been reckoned that a single aphis may give rise in one summer to a *quintillion* (1,000,000,000,000,000,000,000,000,000,000) of little ones!

After this we shall not be surprised that the plant-lice have taken possession of so many of the green things in the world, and the only wonder is that they have not destroyed them all. This they would certainly have done if it were not for their enemies; but the birds delight in them as dainty morsels, beetles and earwigs devour them, and flies lay eggs in their bodies, while the lady-bird eats no other food; and the blind grub of a fly (*Musca aphidivora*) glides about on the stems (see *g*, Fig. 69), seizing the aphides in his mouth, sucking out their juicy bodies, and dropping the empty skins by hundreds.

And now if once your eyes are open to see these tiny juice-suckers, you may find numbers of different kinds living their little lives in the fresh country air. Have you never picked up an apple-leaf or elm-leaf covered with something looking like tufts of white cotton, so sticky that you cannot clear your fingers of it? If so, look carefully at it next time you find it, and under each white tuft you will see an insect struggling along which is like a rose-aphis, only without the little tubes on its back. In fact this fluffy stuff is a kind of wax which oozes out with the sweet liquid all over the body of the insect, protecting it from the sun and from enemies as it feeds, and making it look like a lady in a feathery white ball dress. Some species[*] of these fluff-covered aphides fasten on to the stems of apple-trees, and have been known entirely to destroy them. Then, again, there are others which eat their way into the leaves of trees, making rosy bladders upon them, while others attack the wheat or the hops. In mild seasons, when these insects increase rapidly, they have been known to destroy a whole hop-harvest.

Nor are the aphides the only plant-suckers. Look at the bushes in summer and you cannot fail to see little clusters of froth here and there (Fig. 70), known as cuckoo-spit, because they first appear at the time when the cuckoo sings. Move this froth carefully, and in the middle of it you will find the cuckoo-spit

[*] *Myzoxyle mali* and *Eriosoma lanigerum*.

grub (c) which has given out this froth in bubbles from its tail, to shelter it from the burning sun and hide it from the birds as it sits and sucks the sap. But in the autumn the "spit" if not dried up will be empty, for the insect after losing its last covering-skin

Fig. 70.

C, Cuckoo-spit insect* coming out of the froth. F, The same insect when its wings are grown, and it is known as the Frog-hopper. P, Red Cabbage-Bug.†

has come out with wings—a little brown frog-hopper (F, Fig. 70) which pats down upon your hand, and is gone again in a moment before you have time to examine the wonderful beauty of its wings. Then, again, there are the numberless little scaly plant-lice which spend their lives flattened against the stalks

* *Aphrophora spumaria.* † *Pentatoma ornata.*

and branches of shrubs, looking like little lumps upon the stem, so that while the cuckoo-spit is protected by its froth the scaly plant-lice often escape by their likeness to the colour of the shrub on which they are. It is to these animals that the beautiful cochineal insect of Mexico belongs, and also the lac insect of India, which gives out the red lac used in the manufacture of sealing-wax.

All these are plant-suckers, and rarely move any distance from their home; but there is another group which has learnt to run actively in search of food, of which some suck the juice of plants, while others have made use of their sharp lancet mouths to steal the blood of animals. These active suckers are the air-bugs and water-bugs; and though we dislike the name because one ugly wingless species haunts our own rooms when we do not keep them clean, yet many are very beautiful creatures. Look among the cabbages in the garden, and you will scarcely fail to find a pretty little red and black bug (P, Fig. 70) running over them, and piercing the leaves for their juice; while a grey one with black, red, and yellow spots, is busy at work on the raspberry fruit. If you touch or frighten these plant-suckers, the disagreeable smell which they will give out from their bodies will suggest to you at once that they belong to the bug family.

Then there are those curious thin-bodied insects which skim over the ponds in the summer, actually running on the top of the water, for which reason they are called "water-measurers" (*m*, Fig. 71), because their legs when stretched out seem to measure off the water as they go along. These ghost-like looking insects are water-bugs, whose mouths are made for sucking like the land-bugs, and woe betide the little water-flies which come in their way. Protected themselves from the water by a thick coating of plush under their bodies, they glide along silently and rapidly, and seizing their prey hold it in their

Water-Bugs.
M, The Water-Measurer.* B, The Water-Boatman.†

fore-legs while they suck out the juices. Thus these active little creatures have learnt to find food on the water which their land relations cannot reach, while a still bolder race, the water-boatmen (B, Fig. 71), which lay their eggs on the leaves of the water-plants, dive into the water carrying their air with them, and feed upon the tadpoles and watergrubs below.

* Gerris. † *Notonecta glauca.*

In any pond in the summer time you may see these agile insects rowing themselves along, under-side uppermost, by their two long hind legs, and poking their tails out of the water to take in air under their wings. Then as they row themselves down again you can see bubble after bubble escaping from the tail till they come up for more. You might think they were water-beetles, but their strong wings which they use to leave their ponds by night are all four made use of in flying, whereas in beetles the two front wings are only stiff covers to preserve the delicate ones underneath. Moreover, these bugs have the same lancet mouths as the plant-lice, with which they pierce the skins of the tadpoles, soft grubs, and other water animals near the bottom of the pond and suck out their bodies, leaving nothing but the empty film-like skins. They are most voracious animals and will attack even small fish. Their habit of moving upside down is of great use to them in feeding, for many of their victims have hard backs, so the water-boatmen dive down and come up *under* their prey, thus attacking them on their soft side.

Thus we see that the sucking lice and bugs with their beak-like mouths have bored their way into all kinds of living food; into plants, shrubs, and trees in all countries, and into the flesh of men and animals both on the land and in the water, for many of the lice on birds and other creatures belong to this class.*

And yet in spite of their numbers they are not nearly so destructive as the next class of insects we

---

* Called *Hemiptera*, or half-winged (*hemi*, half; *pteron*, wing).

must speak of—namely, the straight-winged insects,* the grasshoppers, crickets, locusts, and cockroaches, which are not content with sucking, but tear and devour the grass and leaves with their strong jaws. These greedy devourers are a very ancient race of land-insects, and in fact if we attached importance to pedigrees they should have come first in order; for at the time when the piece of coal you put on your fire to-day, and which has been lying for untold ages in the earth, was part of a living forest, grasshoppers, crickets, and cockroaches were already creeping, leaping, and chirping in the dense jungle of ferns and reeds, where they left the remains of their bodies among the decaying plants, so that we find traces of them now in our coal-mines. From that time till now they have struggled on, and while the crickets have learnt to burrow long tunnels underground to hide themselves in, and have homes in the cracks of walls and in company with the cockroaches in the nooks by our firesides, the locusts and the grasshoppers have contented themselves with the open fields and protection of the trees.

> When all the birds are faint with the hot sun,
> And hide in cooling trees, a voice will run
> From hedge to hedge about the new-mown mead—
> That is the grasshopper's: he takes the lead
> In summer luxury—he has never done
> With his delight, for when tired out with fun
> He rests at ease beneath some pleasant weed.

Let us look closely at him (Fig. 72), for he is so large that he will help us to understand the general structure of a six-legged insect better than we could have seen it in the tiny bugs.

* Orthoptera (*orthos*, straight; *pteron*, wing).

INSECT SUCKERS AND BITERS.

We see at once under his wings that he has a ringed body like the prawn, but his head is separate from his shoulders, and carries only one pair of antennæ, while the prawn has two. His mouth, with its two little palpi sticking out, is strong and powerful, and is made of six parts—an upper and lower lip, with a pair of biting jaws within them which

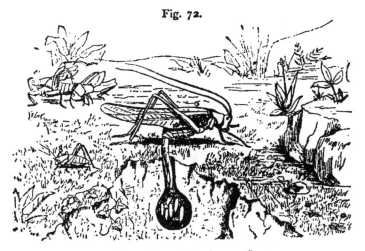

Fig. 72.

Large Green Grasshopper.*

g, Very young grasshopper; *g'*, the same older struggling out of its skin. G, Full-grown female grasshopper laying her eggs in the earth. *o*, Her ovipositor or egg-laying sheath; *s*, one of the *spiracles* or breathing-holes.

move to and fro sideways to hold and cut their food, and within these again are a pair of chewing jaws to masticate it. These six parts, however much they may be altered, we shall find in all six-legged insects, and whether for piercing, sucking, or gnawing they are powerful implements.

* *Gryllus viridissimus.*

Next behind the head of the grasshopper comes his chest or *thorax*, formed of three rings, each of which bears a pair of legs, and the two hinder ones two pairs of wings, while the shoulders are covered by a shield. Behind this again comes the ringed abdomen which he can curve and bend at will. Now, on looking carefully along the sides of these rings you will see a dark spot (*s*) on each, and a magnifying glass will show that this spot is a round plate with a slit in the middle (*s*, Fig. 73). These are the breathing-holes or *spiracles* of the grasshopper, and if you were to cut his body open under water you would see hundreds of minute glistening tubes called *tracheæ* (T, Fig. 73), running in all directions and ending in larger tubes, each of which is joined to one of these breathing-holes. These tubes are formed of two layers, between which is wound round and round a stiff wiry thread (see Fig. 73), which keeps the tube in shape, just like the spiral wire which they put into india-rubber tubing. The glistening is caused by the air which has been taken in at the breathing holes, and fills all the tubes.

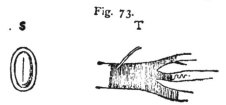

Fig. 73.

S, Spiracle or breathing-plate (*s*, Fig. 72), with the slit in the centre which opens to take in air.

T, Part of the breathing-tubes or *tracheæ*, showing the spiral thread which keeps the tubes in their round shape.

If instead of cutting open the grasshopper you keep him alive under a glass, you may watch his abdomen moving up and down and pumping the air in at the slits, and that air will pass all through his body along these infinitely fine branching canals. Here we have the secret why insects leap, and fly, and run so easily. Think how beautifully light a body must be which instead of containing solid flesh is full of channels of air. Lyonnet counted 1572 tubes in a caterpillar's body, and even then left many smaller ones unnoticed; while some insects, such as the bee and the grasshopper, have not only air-tubes but actual bladders of air filling large spaces in the body. Nor is it only lightness which insects gain by this network of air-tubes, for their blood being bathed in air is always full of oxygen, and therefore active and vigorous, supplying their nerves and muscles with strength quite beyond what we should expect for their size, and helping us to understand why they have been so successful in the battle of life.

So the grasshopper with his large fixed eyes with many hundred windows in them (for structure see p. 224), his delicate feelers, his strong jaws, his long muscular hind legs, and his light body, is an active, powerful insect, and an especially greedy feeder. Indeed he has between his throat and his stomach an apparatus called a gizzard, with more than 200 teeth in it, for grinding the food which he has stripped off the bushes and meadows with his cutting jaws. Even our little grasshoppers in the meadows and the large green grasshopper living in the trees devour greedily all that comes in their way, but their ravages are as nothing compared with those of the large migratory locusts with short antennæ, which multiply at an incredible rate in favourable seasons, and move in swarms over the south of Europe, darkening the sky for miles and devouring every green thing upon their road—

> Onward they come, a dark continuous cloud
> Of congregated myriads, numberless.
> The rushing of whose wings is as the sound
> Of a broad river headlong in its course,
> Plunged from a mountain summit, or the roar
> Of a wild ocean in the autumn storm,
> Shattering its billows on a shore of rocks;

and as they move they leave desolation behind them. Every leaf on tree and bush, every blade of grass, every ear of corn, vanishes under their attacks. In 1866 in Algeria they not only destroyed the vegetables, fig-trees, vines, and olives, but fell into the canals and brooks in such numbers, that the stench became horrible, and the French troops were called out to destroy them and collect their bodies in heaps to burn them; and similar locust plagues cause great devastation in America.

These locusts put their eggs into the earth and cover them up, and the young locusts come out ready at once to begin eating, and exactly like the mother only without wings, which appear later when the last loose skin is cast off. Our little green grasshoppers are locusts with short antennæ, and drop their eggs in this way, but crickets and true grasshoppers, of which our large green grasshopper (Fig. 72) is one, have a very curious and interesting implement for laying their eggs in the earth safe from harm. The mother has a pointed egg-laying sheath or ovipositor (*o*, Fig. 72), at the end of her body, made of several thin plates touching at the edges, and with this she bores a hole in the ground, and then opening the sheath drops egg after egg in to lie till next spring, when they are hatched, if they escape being devoured by underground creatures. Then as each mother has perhaps laid between two or three hundred eggs, it is not surprising that the ground and bushes are covered with tiny grasshoppers leaping and feeding, but without wings. At this time they will all be silent, and as they go on growing will cast skin after skin, till, when the sixth skin is being thrown off (*g*, Fig. 72), their wings appear; and then the young male grasshoppers begin to rub their front leathery wings sharply against each other, so that from their base, where they have a talc-like plate with strong ridges upon it, that shrill cry arises by which they call to their friends.

The crickets, on the other hand, will not be seen in the daytime, for they hide in holes in the ground till night falls, and then come out for food and enjoyment. The only way to entice one out by day is to tickle the hole with a straw, when they will seize it, and so can be pulled out.

In the same way the house cricket hides in its hole behind the oven, where it first came from the egg, and only ventures out to leap and fly about the kitchen at night, when it steals the bread-crumbs and flour, and sips the milk and beer. Often it will begin its chirp long before it comes out, and

> "On a lone winter's evening, when the frost
> Has wrought a silence, from the stove there thrills
> The cricket's song," . . .

for the warmth keeps him awake and alive. But if you leave the room through the winter without fire, then he will sleep in the cracks of the chimney till the warm weather comes back.

All these are leaping "straight-winged" insects, and, like the plant-lice, they feed on vegetable food. But the cockroach, which is not a "black beetle," as people call it, nor a beetle at all,* but one of the

Fig. 74.

Cockroaches.

f, Female with aborted wings carrying her case of eggs, *c* ;
m, male with wings flying ; *y*, young cockroach.

straight-winged insects, has its legs formed for running instead of leaping, and eats all kinds of food,

* I would beg every reader of this book to begin to try always to give the cockroach its true name, for though the false name may be long in disappearing, the confusion might be gradually got rid of if every child used the right word.

whether animal or vegetable, not even sparing our dish-cloths, if they have any grease upon them. Though we have most of us had cockroaches at one time or another in our houses, few people know anything of their history ; of the fact that the mother has only imperfect wings, while the father can fly about ; or that when a cockroach changes its skin it comes out white and soft, and is some time before it regains its dark reddish-brown colour. Nor is it likely that many people will have found the curious little horny cases of eggs (c, Fig. 74), shaped something like a bean, and divided inside into separate compartments, which the mother carries about with her, half out of her body, till the eggs are nearly hatched, when she hides it in the cracks of the boards and mortar of the ovens. These cases contain about sixteen eggs, ranged neatly in two rows, and the edges of the case are strongly cemented together. As soon as the eggs are hatched, the young cockroaches give out a fluid which loosens the cement, and they come out into the world small, wriggling creatures (y, Fig. 74), with all the rings of their body conspicuous, because their wings are not yet grown.

But above all, few people probably would give cockroaches credit for being intelligent animals, and yet an escape of cockroaches which happened in the house of a friend of mine shows them to be more clever than is generally supposed. The house being infested with these animals, the cook laid a trap to catch them, made of a box with two strips of glass sloping inwards from the sides, and it happened that the edges were only about an inch from the bottom of the inside of the box. In this trap she caught many cockroaches, for after getting in they could not mount on the glass again to get out. But one evening, having noticed that the trap was nearly full before she went to bed, she was surprised in the morning to find that all the bait was eaten, and every single insect had escaped. This happened several nights, and at last she resolved to watch. On doing so she saw one of the larger cockroaches stand upon his tail, and so reach up with his front feet to the edge of the glass, and then all the other cockroaches ran up his back out of the box, he dragging himself up last and escaping with the rest. In the open country cockroaches have many enemies which keep them in check; birds and hedgehogs devour them, and bees, ants, and wasps, especially the sand-wasps, hunt them down; but in our houses nothing but cleanliness and killing every one we meet with can rid us of this terrible pest.

And now we must pass over the other straight-winged insects: the Earwig, which uses its pincers to fold its wings neatly under its wing cases and watches over its eggs with a mother's care; the Mantis or snatching insect, which in warm countries creeps along the branches of trees with its forelegs up as if praying, but really in readiness to snatch any passing fly or insect; and the Leaf and Stick insects, which feed on green leaves, and are protected from the birds by looking so exactly like the leaves and stems of the trees they crawl upon, that you may touch them without dreaming that they are living animals. All these are wonderful examples of the tricks which life has taught to her children for protection and attack, but, with the exception of the earwig, the creatures employing them belong to other countries than ours, and we cannot dwell upon them, for we must turn to another group, the "nerve-winged insects,"* which we meet with every day, and whose history is perhaps one of the strangest among insects.

In the time of those ancient coal-forests of which we have spoken, when the grasshoppers and cockroaches lived upon the land, another race of insects, belonging half to the water and half to the air, were spending their youth in the ponds and marshes, and hovering over them in their riper age. These were the ancestors of our May-flies and dragon-flies, and from that day to this they have kept up this strange existence, hunting and chasing their prey at the bottom of ponds until the time comes for their wings to grow, and then climbing up the water-plants, and bursting forth into glorious winged animals, which

> "To the sun their insect wings unfold,
> Waft on the breeze, or sink in clouds of gold;
> Transparent forms too fine for mortal sight,
> Their fluid bodies half dissolved in light."

Every one who has been on a warm summer's day near the borders of a lake or pond, must have seen those delicate and fragile flies called May-flies (*mf*, Fig. 75), which dance in the sunshine, flag as the sun goes down, and die in the night. They are not difficult to know with their widespread unequal wings, their short delicate antennæ, and their bodies ending in three long fine bristles; and they do nothing but rise

---

* Called *Neuroptera* or "Nerve-winged," on account of the network of veins in the wings.

and fall in and out in a mazy dance; for they have but a few hours to live, and their mouths are too soft for them to take food. In fact, the whole end and aim of their winged life is to form and lay eggs in

Fig. 75.

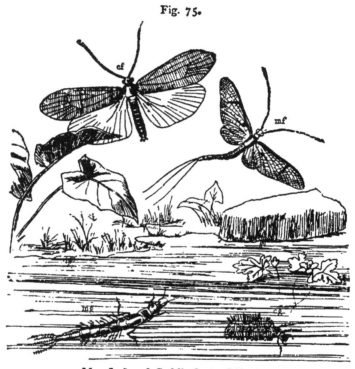

May-fly * and Caddis-fly.† Life size.
mg, May-grub, with its fringe of breathing gills; *mf*, May-fly;
cg, caddis-grub in its case; *cf*, caddis-fly.

little packets on the water to hatch into future young ones. And yet they are not really mere "creatures of a day" as they have been called, for before they

* Ephemera. † Phryganea.

INSECT SUCKERS AND BITERS.

obtained their wings they lived for nearly two years at the bottom of the pond over which they now fly.

Their real life is that of a water-insect (*mg*), which as soon as the eggs are hatched, dives to the bottom of the pond and burrows in the ground or under stones, and feeds upon all passing insects, seizing them with strong spiny jaws, and devouring them greedily. At this time the May grub does not breathe through holes in the side, but has its body fringed with gills or delicate folds of skin which take in air out of the water, and there is nothing in its appearance to lead any one to believe that it could ever live in the air. But as it grows up and loses one loose skin after another, the rudiments of wings are seen through the transparent covering, and then the end of its life is beginning. It creeps out upon some plant or stone overhanging the water, the skin cracks down the back, and the flying insect comes out with its wings perfectly visible. Still it cannot use them, for a fine film covers the whole body, and it is only a few hours after, when this has dried and split, that the perfect May-fly soars away an air-breathing insect, to lay its eggs and die.

We need not then pity these frail, delicate ephemera on account of their short life, for they have had a long and merry one feeding in the pond below, and when we see them, they are taking their last enjoyment before night falls.

The same thing is true of the caddis-flies or watermoths (*cf*, Fig. 75), which anglers use as bait, for they too cannot feed after they get their wings; but their life in the pool below has been rather different. Their tail is soft like that of the hermit-crab, and they need to hide it in some strong covering. And for this purpose they build themselves tubes of silk, into which they weave pieces of wood and grass, or of sand and stone, and even sometimes shells with living creatures in them; and dragging these tubes about with them, they put out their strong head and shoulders, and feed on plants and insects. Then when the time comes for their change they draw back into their case, and closing it with a grating of silk at each end, so that water can get in while enemies are shut out, they lie still for a fortnight, like a caterpillar in its chrysalis, and then swim out and creeping up some plant burst their covering skin, and hover over the pond, or rest upon the bushes till their eggs are laid, and they die.

Thus, the May-fly and Caddis-fly live chiefly in the water, finding plenty of food during all their growing time, while they have but a short glimpse of the pleasures of the air. Not so, however, the gorgeous-winged Dragon-fly. He manages to make the most of both worlds, and, whether he is crawling in the water below, or flying in the air above, is one of the most voracious and bloodthirsty of insects.

Even when he is a water-grub, though he moves very slowly, yet the quickest of insects cannot escape from him, for he has a peculiar under lip (*m*, Fig. 76), very long and with two sharp hooks at its broad end, with which he seizes them. This lip folds back by a kind of hinge, and is called a mask, because it covers the lower part of his face, and makes him look an innocent and harmless creature. The moment an insect comes by, the lip is shot out, and the pincers grasp the prey, throwing it into his mouth

INSECT SUCKERS AND BITERS. 223

as the mask again closes. So gliding stealthily along the bottom, the greedy creature seizes all that comes in his way—grubs, worms, water-slugs, and even small fish, are all attacked by him; and though he

Fig. 76.

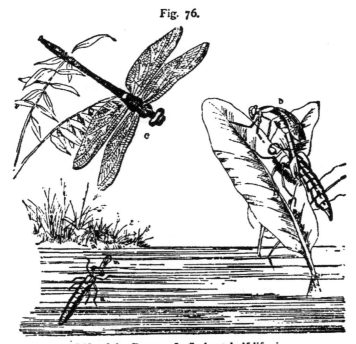

Life of the Dragon-fly,* about half life-size.

a, Grub living in the water; *m*, mask or long lower lip with which it seizes its prey; *b*, dragon-fly creeping out of its last grub skin; *c*, perfect dragon-fly on the wing.

is sometimes devoured in his turn by other animals, yet he often escapes, for he breathes by taking in water at his tail, and when he wishes to get out of

* Libellula.

the way, he shoots this water out, and drives himself along much as we saw the octopus do, only that the dragon-fly grub goes forwards instead of backwards.

So in a year he grows big and strong, and short wings begin to appear under his skin. Then he crawls listlessly to the top of a plant, and there dragging himself out of his covering, he gradually expands his large gauze wings filled with delicate air channels, and shaking free his sharp-clawed feet, is at once ready for new victims. His large gleaming eyes with their thousands of windows (often 12,000, each with its own lens and cone and rod, see Fig. 77) espy a butterfly or a moth, and in an instant he is pursuing it, flying upwards, downwards, and sideways, without turning, by means of the peculiar muscles by which his wings work upon his bulky shoulders, while his long body serves as a rudder. And when he has caught his prey, he tears it savagely with his horny jaws, scattering its shattered wings upon the ground. Much as we admire the beautiful colours and magnificent wings of the dragon-fly, we must admit that when he reckons back his ancestors through the dim ages to those distant coal-forests, he must look along a line of the most greedy and cruel marauders of the insect world.

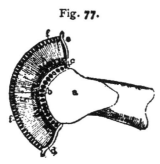

Fig. 77.

Section of an Insect's Eye.

a, Nerve-mass of the eye; *b*, nerves springing from it; *c*, the retina; *d*, thread-like rods by which the picture is formed; *e*, glass-like cones; *ff*, lens-like facets, of which each eye may have from 10,000 to 20,000.

Nor has he been content with ravaging the water and the air only, for one of his very near relations, the ant-lion (*Myrmeleon formicarius*)—which is to be found in France and most warm countries, and which when it has its wings might be mistaken for a dragon-fly—lives its early life in dry sand, in which it twists round and round, till it has made a funnel-shaped hole at the bottom of which it lies (see plate, p. 135). This it does near an ants' nest, and when an ant running on the edge of the funnel slips in, the ant-lion flings sand upon it so that it tumbles to the bottom, and he can devour it. Thus in the water, on the land, and in the air, the dragon-flies have a good time of it, if they can only escape the swallows and other quick-flying birds, which pounce down upon them, and the scorpion-flies, which, though much smaller than themselves, sting them to death.

And now we come to the most interesting of all the nerve-winged animals, the Termites or white-ants, the plagues of India and Africa. Every one has heard of these destructive creatures, which feed so cunningly out of sight, eating their way from the ground beneath, up the middle of posts, beams, woodwork, and furniture; and even sometimes propping up with hardened mud and slime the buildings they are eating away, so that no one finds them out till all at once some part falls down and exposes the rottenness within. They are so clever, and have so many habits like true ants that they have been called by their name, and most people think that they are their near relations. But this is not so; on the contrary, while the ants stand at the head of the most highly developed insects, and have a helpless infancy, and a true chrysalis state before they are fit for work, the termites have come along quite a different line, and belong to those ancient

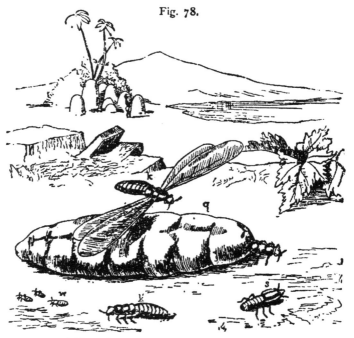

Fig. 78.

African Termites * taken out of their home.
k, Winged king ; *k'*, same after losing his wings ; *q*, queen ; *w*, worker ; *s*, soldier.   All natural size.

nerve-winged insects, which work from their earliest age, and have no time of rest to prepare for their grown-up life. So when we find these "white-ants" living together, and building houses, and having workmen and soldiers besides the true king and

* *Termes bellicosus.*

queen of the nation, we see that they must have learned these habits quite independently of the true ants, with which they have nothing to do.

A strange and wonderful thing a termite community is. Perhaps in India you find one day that the sill of your window or the post of your door is rotten, and then when you begin to cut it you find it completely hollowed out into little chambers, the wood being eaten away. At first these chambers are empty, but as you go on, you find small soft white insects with six feet running hither and thither. These insects are quite blind and wingless, and always work in the dark. Even if they come out on the surface of the wall or the ground, they cement wood-dust together, or carry up clay to make a tunnel, under cover of which they travel up and down. In a single night a tube may appear all up your wall, which is a termite tunnel, built to enable the insects to reach some fresh store of woodwork. If you watch you may see the tunnel grow at its open end, as one little white grub after another comes to the opening and laying on its little bit of mudwork, goes back to make room for the next. Mingled with these workers are much larger insects, also blind and wingless, with huge heads and jaws shaped like jagged stilettoes. These are the soldiers; they do not work but defend the labourers, hanging on to any enemy with their sharp pincers, and allowing themselves to be torn to pieces rather than give way.

And now as you penetrate farther and farther through the woodwork and probably down into the ground below, the small chambers begin to be filled with little white eggs, and with snow-white young workers and soldiers, which feed on a kind of mouldy fungus growing on the walls of their rooms. At last in the midst of these you come upon a large cell with a long soft whitish-brown lump in it as big as your finger, and looking something like an uncooked sausage (Figs. 78 and 79). At first you would think this was a mere bag, but looking at one end you would see three rings, each with a pair of legs, and a head (*ht*, Fig. 79) with eyes and feelers and

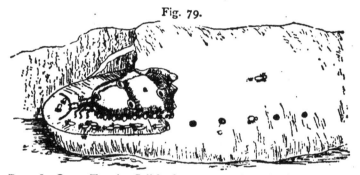

Fig. 79.

Part of a Queen Termite Cell broken open to show the Queen within.
Smeathman.

ht, Head and thorax, *a*, abdomen of the queen; *o*, only real openings in the cell when it is perfect. The workers pass through these. *w w*, Workers.

mouth. This white lump, then, is part of a living creature; it is the abdomen of a termite queen swollen to nearly 2000 times its natural size and full of eggs. There she lies with her husband (*k*, Fig. 78), who is much bigger even than the soldiers, but nothing as compared to his queen, crouching by her side; and while the working termites feed her and caress her, she goes on laying eggs incessantly, about 60 a minute. These the workers carry away as fast as she lays them, and store them in the nursery-chambers around; but the holes through which they pass out of the queen's cell (*o*, Fig. 79) are far too small for either of the royal pair to escape.

The history of this curious community has been as follows: About two years ago, before the heavy autumn rains began, this king and queen, with thousands of others, were born from the eggs of another huge queen, and when they had cast their skin, they came out each with four gauzy wings, and flew into the country, or often into the houses if their nest was in a town. But their bodies were heavy and their wings weak, and so they soon fell to the ground, where nearly all their companions were eaten by birds and ants and other creatures. They, however, chanced to escape, and losing their wings were found by a party of working termites. At once these active wingless workers carried the royal pair into their tunnel, and built a clay chamber round them, with only small openings in it (Fig. 79), not large enough for them to get out. There they fed them carefully, and by and by the abdomen of the queen began to swell, and from that time her whole mission was to lay eggs. As her size increased, the workers enlarged her chamber, and meanwhile were toiling busily making nurseries for the eggs, and storehouses for the shavings of wood and masses of vegetable gum, which they collected by burrowing in every direction through living trees, or beams, or woodwork of any kind.

You may imagine how many nurseries must be built, besides new rooms for grown-up workers, if 80,000 eggs are laid every day; and besides these nurseries there are innumerable galleries and passages, which are all so arranged that air passes through the whole building. The work goes on in perfect order; some tend the queen, and all show her the utmost attention and affection; some store the eggs; some look after the young; while others enlarge the building or tunnel for long distances underground to get food. And all this is done by blind workers in pitch darkness, with a regularity and precision which is most marvellous, and can probably only be accounted for by the supposition that their antennæ are far more delicate and useful implements than we can as yet understand.

There are many species of these Termites. Some live in buildings, and the town of La Rochelle in France where they have probably been brought from the West Indies, has been sadly damaged by them. Others in Africa build enormous mounds of clay and earth (see Fig. 78), as much as from twelve to twenty feet high, and so strong that the buffaloes stand upon them to look over the plains; and inside these are innumerable galleries and floors of storehouses and nurseries. If an attempt is made to destroy these mounds the soldier termites, which are about one in a hundred as compared to the workers, swarm out and fall upon the enemy, while the workers begin at once to repair the damage. Other species build nests in tall trees, driven there, no doubt, to escape from the true ants, which, having hard bodies, can attack the soft termites and destroy them easily. Lastly, there are some which come above ground and march from place to place like a regular army; and these are the most remarkable, for not only have the soldiers and workers eyes so that they see their way, but Smeathman, who studied them in Africa, saw the workers marching in regular streams or columns twelve or fifteen abreast, guarded by the soldiers, while sentinels were placed on plants along the road. These soldiers struck the leaves every now and then with their jaws, making a ticking noise, to which the workers answered with a hiss and then quickened their pace.

And here we must end our history. There is probably nothing more curious in the whole insect world than the termite communities; for these children of Life have learnt lessons far above any of their near relations, while the necessity of preparing for and tending the eggs which the queens lay at such a prodigious rate, makes the whole nest a constant scene of activity and contrivance. The study of their habits and customs is one of the greatest possible interest, but here we must content ourselves with a mere general glimpse, and with establishing firmly in our minds the fact that the "white ants" of India and Africa are quite a different race, and belong to a totally different order of insects, from their darker namesakes in England. They belong to the nerve-winged insects, and together with all those included in this chapter they are born in the same shape as their parents, only without wings.

For we shall notice that the aphides and the bugs, the grasshoppers and cockroaches, the mayflies and dragon-flies, as well as the termites, change their coats as they grow too small for them, creeping out of their skins many times in their lives. But they do not change their bodies, as we shall see in the next chapter is done by the caterpillars and butterflies. Neither have they, with the exception of the caddis-flies, any time of trance as the caterpillar has in its chrysalis. They are active from birth to death, and though, when their time for laying eggs comes, they put forth wings to carry them to their mates and to suitable spots for laying, still they have not yet fallen upon the expedient of taking a time of rest and forming a new and beautiful body.

CHAPTER XI.

INSECT SIPPERS AND GNAWERS WHICH REMODEL THEIR BODIES WITHIN THEIR COATS.

> And many an antenatal tomb,
> Where butterflies dream of the life to come,
> She left clinging round the smooth and dark
> Edge of the odorous cedar bark.
> THE SENSITIVE PLANT.

AMONG all the strange and puzzling facts in the history of living things there is perhaps none which has attracted so much attention as the complete metamorphosis or change in the bodies of insects, by means of which a creature begins its life in one form, then hangs itself up in a hard skin or a silken shroud, or buries itself out of sight, and comes out at last so totally different in appearance, that if we had not watched it passing from the one shape to the other we could never have believed it to be the same creature.

Who would believe at first sight that a butterfly or moth had once been a creeping caterpillar; or that the hurrying busy beetle and the active fly had burrowed as maggots or grubs in decaying matter, or in the trunks of trees, or in the fruits and flowers of plants ; or still less that each active hopping flea once rolled about helplessly as a little hairy grub and span a tiny silk cocoon, in which its present body was formed?

To those who have only paid attention to the higher animals, such as birds, fishes, and quadrupeds, which, when they are born have already assumed a fairly settled form, this springing up of one being out of the husk of another apparently quite unlike it, seems strange and unnatural ; and in olden times all kinds of fanciful ideas were connected with the metamorphosis of insects. But if we begin, as we have now done, by Life's simplest children, and see how in each successive group the necessity for making the best of everything causes many creatures to alter their form and habits at different periods of their lives, then these curious changes in insects have a real meaning, and we can set to work patiently in the hope of discovering what advantage they are to the creature which undergoes them.

Thus, for example, we have already found the sponge beginning as a swimming animal (see p. 38), and then drawing in its lashes and settling down to build a solid skeleton clothed with a colony of cells ; while the lasso-throwers begin by swimming, go on as stationary, branching and budding animals, and end by throwing off egg-bearing jelly-bells quite as unlike the animal tree as a butterfly is unlike a caterpillar. In the star-fish and sea-urchins the transformation scene is still more curious, for the jelly-animal is swimming about and feeding with its special mouth and stomach (see p. 78), while a second and differently shaped form, with a mouth and stomach and feet of its own, is growing up inside; yet both these beings are part of one single creature, and when the form within is ready to get its own living, it swallows its earlier self and goes its way. So too the headless mollusca, such as the oyster and cockle, swim about in their young life and have eyes which they lose when they settle down, while the crab undergoes such complete changes (see Fig. 59, p. 167) that no one would recognise parent and child if they saw them together.

We learn then that it is not the exception, but in many cases the rule for a creature to take on different forms at different times of its life; and the chief novelty in the metamorphosis of insects turns out to be that they have learnt to do one thing at a time, and after passing their early life in incessant feeding and storing up of material for a more perfect body, they retire from the world to spend all their energy in building up those new and beautiful bodies which we admire so much in the lovely painted butterfly or the gorgeous metallic-winged beetle.

Nor shall we wonder that this quiet is necessary when we understand the marvellous change which takes place in them. The cockroach and the cricket only gain wings by their last change of skin, and though the May-fly alters its apparatus for breathing so as to be able to live in the air, still the greater part of its body remains the same. But the caterpillar and the grub have actually to remodel every part of their bodies in order to become the butterfly or the beetle, and we can scarcely say that any portion of them remains as it was, except that mysterious life-power which brings to them from past generations the experience to guide them in their development and their work. Yet so true is this experience, so well has the lesson been learnt by the countless ancestors which have gone before, that among the thousands of different kinds of grub, and maggot, and caterpillar, each follows its own peculiar road as its forefathers have done before it, and wrapping itself in its own special form of covering, goes through its curious change, and comes out as fly, butterfly, moth, or beetle, with those weapons and ornaments which it needs for the rest of its existence.

Of all the marvels of life, surely this is one of the most marvellous, and why or how each one puts on its peculiar dress we can scarcely ever hope to know. But we may gain some slight idea of the general process by which a creeping worm is changed into a winged insect; and to do this let us sketch out the life of the common Tortoise-shell butterfly which Mr Newport watched through all its changes nearly fifty years ago.

It is under the fresh green shoots of the common nettle that the tortoise-shell butterfly-mother lays her cluster of eggs in the early summer, after she has crept out of the crack in the wall or paling, where she had remained hidden for her long winter sleep. Coming out with worn wings and faded colours, she enjoys a short flight in the sunshine with others of her kind, and then leaves her eggs where the young will find food, and goes her way to die. And in a week or two, when these eggs are hatched, there creep

INSECT SIPPERS AND GNAWERS. 237

out little yellowish-grey caterpillars, and these give out at once a fine silken thread from their under lip and spin a slight web over a leaf. This done, they crawl away in company and travel from leaf to leaf, feeding busily all day but always returning to their silken tent at night. They have no thought beyond eating as they move along on their ten cushion feet, two at the tail and four under the abdomen (*cf*, Figs. 80 and 81), which are not true feet at all but foldings in the skin, each bearing a circle of spines which

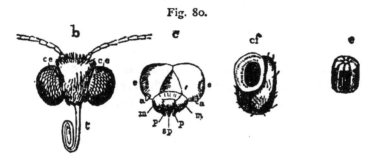

Fig. 80.

b, Butterfly's head ; *t*, the trunk ; *c e*, compound eyes ; *c*, caterpillar's head; *a a*, antennæ ; *m m*, mandibles; *p p*, palpi of the jaws ; *sp*, spinning-tube ; *cf*, cushion-foot seen from underneath, showing the circlet of spines ; *e*, egg of the tortoise-shell butterfly. The true size is about that of a rape seed.

help it to cling to the twigs. Their six real jointed feet (*f*, Fig. 81, p. 239) near the head, they use both for walking and for grasping the leaf, while they cut it with their horny mandibles or outer jaws (*m*, Fig. 80), which work horizontally between their lips, and then pass the pieces on to be chewed by the real jaws within, whose palpi or feelers are seen at *p*.

Thus they "feed and feed alway," guided probably chiefly by touch and taste, though they have tiny simple eyes. They do not need to pause for breath since that is taken in through the holes in their sides (*b*, Fig. 81), and they eat so greedily that after a time their skin becomes too tight for the food they are packing into it. Then they pause and turn pale and remain still for a while, after which each one bends up his back, swells out his rings, and so splits the inconvenient coat along the back. Then drawing out first his head and then his tail, he comes out fresh in colour with every joint and hair in its place, and begins gorging once more. This they do as many as five separate times, and at the end of these changes their new form is already growing within them, for if you cut open a caterpillar just before it casts its last skin, you may see the outline of the wings and antennæ of the future butterfly in a watery state, each in its transparent sheath.

And now they must shut themselves up from the outer world, for each one has to make a sipping mouth instead of a biting one, and to gather up his muscles to make his shoulders strong to bear his wings; and above all he has to draw together the line of nerve knots which in the caterpillar are stretched along his body as we saw them in the leech (p. 143), but which in the butterfly must be concentrated so as to make great central nerve stations in the head, for the use of the large eyes and sensitive antennæ which are coming, and other stations under the shoulders to supply his wings with energy.

So each caterpillar again leaves off eating, and finding some firm spot on the trunk of a tree or a post, or a stem of a plant, makes there a little hillock

INSECT SIPPERS AND GNAWERS. 239

of silk, and clinging to it by his hind feet, lets himself swing head downwards (Fig. 81). Then his head and shoulders begin to swell, the nervous knots within his body to draw together, the air-tubes to expand, and the skin to crack, so that by vigorous efforts he

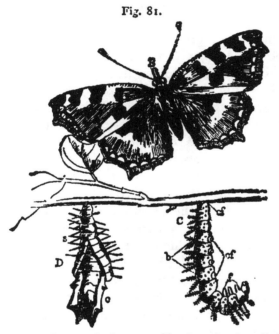

Fig. 81.

C, Caterpillar hanging by its two cushion feet *cf* at the tail, the other eight cushion feet are in the middle of the body. *f*, The six true feet; *b*, breathing holes. D, Chrysalis breaking through the caterpillar skin. B, Perfect tortoise-shell butterfly.

can push his whole body covering back to his tail, where at last it drops off, leaving him hanging by some small hooks at the end of his body. A curious fellow he looks now as each part of the future butterfly is dimly seen in its protecting sheath. His tiny wings, his six true feet, his antennæ, and his sipping trunk, have all begun to form, but are far from complete; and to keep them safe till they are full grown, a clear fluid oozes out and flows over all, hardening into a firm transparent case; and as in some butterflies the reflection of light from the under surface of this case has a golden tint, the name of chrysalis (*chrysos*, gold) has been given to the still and quiet form; but the word *pupa* (doll) is perhaps better, because many have not this golden hue.

It is within this sheath that in about three weeks the butterfly's body is gradually formed, and all the fat which the caterpillar had stored in the spaces of its body is worked up into muscle and nerve and egg-producing parts. At last all is complete; the head, shoulders, and abdomen have taken on their real shape; the delicate tinted scales which cover the wings, and deck them with gorgeous colours,[*] are full grown; the wings themselves, made of two layers of skin between which are the air-tubes and the veins presently to be filled with colourless blood, are all ready; while the little pockets in the body which make the full-grown insect so much lighter than the caterpillar, are waiting to be expanded with air. The nerves begin to send messages to the limbs to move, and the perfect butterfly, splitting its transparent covering, creeps out into the world, slowly but surely inflates its body and wings, and letting them dry in the sun, soars off to sip the flowers and find a mate.

But what a different creature we have here from

[*] Hence the name Lepidoptera (*lepis* a scale, *pteron* a wing), or scale-winged insects.

the creeping, gnawing, spinning caterpillar! The two lips of the caterpillar with the silk-spinning tube (*sp*, Fig. 80) in the lower one, are reduced to mere fragments, the horny mandibles (*m m*), no longer needed now the chief feeding time is past, have almost disappeared, while the two inner jaws are drawn out into long hollow channels and fit together so as to form the delicate tube (*t*) which is to suck honey from the flowers. The tiny eyes (*e*) of the caterpillar, if they are still to be found in the forehead, are quite insignificant when compared with two large, many-windowed eyes (*e e*) which now stand out on each side of the head to warn the rapidly-flying insect of danger from all directions. The tiny stunted antennæ of the caterpillar have become long and delicate. The shoulders, grown firm and strong, carry the six slender legs, and two pairs of wings which are worked by powerful muscles; and over these wings is spread a carpet of beautiful scales, each one fitting by means of a little tube into the wing, and the whole making a brilliant pattern to attract the eyes of the mate which the butterfly now wishes to find. The abdomen has lost on the outside the cushion feet which are no longer needed, while within, the long digestive tube which it had as a caterpillar has become quite small, making room for an apparatus for forming and laying eggs.

And yet though such a total remodelling has taken place, there has been no such thing as death and new life between the caterpillar and the butterfly. Though the chrysalis hung in such a still and death-like form, it was the same living insect, breathing almost imperceptibly, and able to move slightly if touched. Only the life within it, which in the first stage was busy storing up material in the caterpillar, was entirely occupied during this second stage in moulding that material into the future butterfly, which in the third stage as a perfect insect completes the history.

We see then, that one of the great questions in all creatures which remodel their bodies must be, how they can keep themselves from danger in this second stage when they are so helpless. Some go through their changes quickly, and then they are comparatively careless of anything but to choose a secluded spot. Our tortoise-shell butterfly, for example, hangs very insecurely by the slender thread of her chrysalis. But then she is generally but little more than a fortnight or three weeks completing her change, and even when born in the autumn she becomes a butterfly before the winter, and goes to sleep in this form, hiding in the chinks of walls or palings, or in the bark of trees, till the warm spring comes round—unless, indeed, some mild day in January wakes her before her time, when she generally dies as the penalty for mistaking the season.

But the common cabbage butterfly, if born late in the year, often remains as a chrysalis from September to April, and would hang very unsafely exposed to the rough winds if merely attached by the tail. So the caterpillars of this butterfly, as of many others, bind themselves firmly to the paling or wall by a narrow band of silk. If you can catch sight of this caterpillar just when beginning its change, you may see it first make a little tuft of silk (t, Fig. 82), in which it plants its tail, and then turning back its head, pass the silk from its tip across and across the body (*b b*), so that by and by when the skin is shed, the chrysalis remains firmly tied to the paling.

One of these two ways of fastening themselves are followed by almost all the caterpillars of butterflies, except a few which roll themselves in leaves or bind themselves in slender webs. But the moth-

Fig. 82.

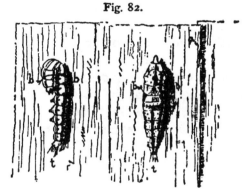

Caterpillar and Chrysalis of Cabbage Butterfly* bound to a paling.
t, Tuft of silk holding the tail ; *b b*, Silken band securing the chrysalis.

caterpillars are much more clever at hiding, and in many ways are more interesting than those of butterflies.

Naturalists are in the habit of dividing the Lepidoptera or scale-winged insects into moths and butterflies, and although there is no real distinction between them, yet in a general way it is not difficult to tell them apart.

A moth, as a rule, lays its wings down upon its

* *Pieris brassica.*

back when at rest, while a butterfly folds them up, back to back against each other, and though some moths copy the butterflies in this, they are not many. Again, most butterflies have their antennæ thick at the tip, while those of moths are more generally fine at the end, and thicker in the middle, and are often beautifully feathered, but this rule also is not without exceptions. Again, the wings of moths are fastened together by a kind of hook, which makes them work much more strongly, and not with the irregular movement which we find in butterflies; lastly, the shoulders of moths are broader than those of most butterflies, and less distinctly divided by a waist from the abdomen. By some of these characters, as well as in many cases by their nocturnal habits, moths may be generally known, although it must be remembered that they are such near relations to the butterflies, that no clear line can be drawn between the two.

But in their habits and devices, the moths far outstrip the butterflies. It is their caterpillars which among the sphinx moths remain motionless for hours on twigs with the head bent down, so as to look like part of the bush; thus escaping the notice of the birds, which would eat them, and of the ichneumon-flies which would lay their eggs on their bodies. And these same caterpillars, when the time of their rest comes, burrow into the ground, and line their home with varnished silk, so that no water can creep in. Here, safe and sound from wet and cold, they cast off their skin for the last time, and lie as pupæ during the long winter, till the warmth of June wakes them into moth life.

It is the caterpillars of moths, too, which spin those silken cocoons which hang from tree or bush, or under walls and palings; homes so delicate, and yet so dry and snug, that the tender pupa lying freely inside them is like a child in its warm bed at night. Any one who has kept silkworms will know how cleverly the caterpillar, bending its head back towards its tail so that its feet are outside, begins its outer egg-shaped layer of silk by moving its head to and fro in some nook or corner, and leaving a bed of fluff within which it spins the coil of finer silk. You may watch the cocoon growing for a time as the caterpillar's head moves round and round in an oval form, leaving its silken trail behind it. But gradually the meshes grow finer and finer, and you can no longer see through them, while still the industrious creature goes on till its head has been round the oval at least three hundred thousand times, and it has made a stout cocoon.

Once safe inside its silken house, it pushes off its caterpillar skin and remains a protected pupa for a fortnight or more. Then, if you have not already robbed it of its silk, the moth, after it has crept out of the pupa skin, must work its way through the cocoon. This it does by giving out a liquid which is contained in a little bladder in its head, and soaking the silken wall so as to separate the threads and make a path for itself to the outer air. But curiously enough it will not attempt to fly far, for the silkworm moth, belonging as it does to a genus already feeble in flight, and having besides been kept in confinement for generation after generation, makes scarcely any use of its wings.

Out of just such cocoons as this, but of coarser make, with a tiny hole left at one end, come the beautiful emperor-moth, the night peacock, and the curious Oak-eggar moth, whose caterpillar sleeps all through the autumn and winter before beginning to feed and spin its cocoon ; while the Burnet moths

Fig. 83.

The six-spot Burnet-Moth.*
c, Caterpillar ; *co*, cocoon ; *m*, perfect moth.

(Fig. 83) often spin very thin cocoons covered with a kind of varnish which makes them as strong as parchment. With a little trouble you may often find the empty cases of these and other moths left on the grass and bushes in July and August, when the insects are fluttering over the gardens and fields. But the

* *Zygæna filipendula.*

INSECT SIPPERS AND GNAWERS. 247

cocoons of the Procession-moths, which climb the oak-trees at night to feed, you will find all enclosed in one large nest of silk, for these caterpillars live in companies in a hanging web, and when they are ready to remodel their bodies, they strengthen the web with their moulted skins, and lie all together, each spinning his tiny cocoon round his body.

Again, there are many caterpillars which have not sufficient silk to spin a whole cocoon, and they have learnt other devices. Thus, some of the sphinx caterpillars make cocoons of dried leaves, woven together and lined with silk; and rolled up in these, they lie under the cover of some stone or bush.

Fig. 84.

Psyche graminella.

g, Front part of the caterpillar with the six true feet; *c*, case of straw, and grass covering the rest of the caterpillar, and in which it will hide as a chrysalis; *m*, perfect moth.

The hairy caterpillars also, many of them use their hairs for the cocoons, binding them together with a little silk; while a group of moths called Psyches (Fig. 84), because they are so small and light, come from caterpillars, which, as soon as they are born, take pieces of straw, or leaf, or grass, and bind them together into cases, in which they live, feeding under cover of the little house, which they enlarge from time to time, and use later on to shelter their chrysalis. It is worth noticing that these caterpillars, living in a case, do not need broad false-feet to clasp the stems, so these are reduced to quite small cushions, with a ring of strong hooks to hold fast to the case. Then there are the leaf-rolling caterpillars, which twist up the margins of leaves, and use their silk to bind them into tubes for their resting-places, while the huge caterpillar of the Goat-moth gnaws its way into the old trunks of willows and elms, and after feeding and tunneling there for three years, creeps just under the bark, and gums together a cocoon of powdered wood lined with soft silk, in which it lies safe and snug till transformed into the large and beautiful moth.

It would require a whole volume to trace out the many devices of the moth-caterpillars to escape their enemies; and to find shelter from wind and weather during their retreat from the world, in some cases of weeks, and in others of many months. But, with the exception of the Goat-moth, all those of which we have spoken feed openly as long as they are caterpillars on the leaves of trees and plants, and have no special means beyond their green or brown colour, or sometimes their nauseous flavour, for eluding their persecutors. It remained for the tiny Leaf-miners to find out the plan of living between the two sides of a leaf, and so eating their way peacefully in covered galleries. These little caterpillars coming out of their eggs on the under side of a rose-leaf, or honeysuckle leaf, bore at once into it, and creep along eating the flesh of the leaf between the two surfaces, till they are full fed, and then they pierce through the upper skin, and creeping out spin those curious little orange cocoons which you may find in the summer clinging to the stems of roses. If you have once looked for the tracks of these tiny insects, you may easily find them showing as pale wavy lines on the honeysuckle and other leaves.

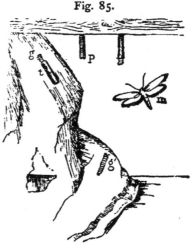

Fig. 85.

The Clothes-Moth.*

g, Grub feeding in its woollen tube *t*; *g'*, naked grub taken out of the tube; *p*, pupa hanging in the tube; *m*, moth.

So you may also trace lines something of the same kind, but more unpleasant in our eyes, on our own woollen clothes which have been laid by for the summer. These are made by caterpillars of the same family as the leafminers, but as there are no covering skins here between which they can lie, the clever little fellows build tubes for themselves (*t*, Fig. 85) out of the wool which they tear off the clothes. They live in these just as the Psyche caterpillars live in the grass tubes, and when they are going to remodel their bodies they close one end of the tube and fasten it to the side of the box or cupboard (*p*, Fig. 85), and then turning themselves with their head to the open end,

* *Tinea tapetzella.*

are ready to come out when they have developed into those little grey moths we know but too well.

And here we must leave the butterflies and moths, without touching upon those moth-caterpillars which live in the water, or those which steal the honey from the bees, or the tiny butterfly-caterpillars which live in the clover and grass, and whose eggs we tread upon as we walk. Each little butterfly or moth which we watch gamboling in the sunshine, or disturb from its sleep in the hedges or on the moss-covered walls, has its own habits and history, its favourite plant on which it feeds and to which its caterpillar feet are often specially adapted, its time for flying and for resting, its special hiding-places for its pupa, and its own lovely markings on the wings, which when open attract its mates, and when closed often shelter it by making it look like the plants upon which it alights; while many moths which fly at night have even a peculiar scent by which they find each other in the dark. And one and all have their object in life—the male butterflies to find a mate, and the mothers to find the plant on which they themselves fed as caterpillars, there to lay their eggs. Moreover, they are unconsciously doing useful work, for as they pass from flower to flower sipping the honey, they carry the pollen-dust on their bodies and fertilise the lovely blossoms which enliven our fields and hedges, and in so doing help to make the seeds which grow up into fresh plants for those which come after them.

But as these delicate children of Life flutter through the world, innumerable dangers meet them on their way—as caterpillars, pupæ, and butterflies, hundreds are destroyed by birds and by other insects, while the pitiless wind and soaking rain of our English summers often batter their tender wings before they can creep under shelter. In this respect they are far worse off than our next group, the Beetles, which are gnawing insects during both the active seasons of their life, and whose front wings are not used in flying, but form those beautiful sheaths called elytra,* which so often make these insects look like brilliant jewels. These elytra in many beetles are very hard and strong, and serve to cover up safely the pair of large transparent hind wings which are used in flying.

There can scarcely be any doubt that the beetles are especially well provided with weapons for fighting the battle of life, for they have not only managed to spread into every country on the globe, but are by far the most numerous of all insects. From the huge Goliath beetle of Africa, five inches long, down to the minute rove-beetles which give such sharp pricks when they fly into our eyes on summer evenings, beetles are of all sizes, and live in almost all conceivable ways. While many feed on plants, others are fierce hunters and even cannibals, devouring each other in the most cruel manner, while a very large number feed on dead and decaying matter and are most useful scavengers, and not a few feed on animals when young, and plants after they awake from their long sleep. For beetles, like butterflies, have three lives—first as grubs or maggots; secondly as helpless pupæ or swaddled insects; and it is only when they come to the third stage that they are true beetles, with wings and the power of laying eggs.

* For this reason they are called sheath-winged insects, or Coleoptera (*koleos* sheath, *pteron* wing).

252 LIFE AND HER CHILDREN.

The cockchafer or May-bug, which blunders up against us as he flies heavily in the night air, began his life underground more than three years ago. His mother, groping down into the earth in the early spring, hid herself there and laid from thirty to forty

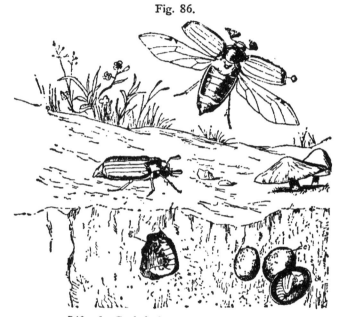

Fig. 86.

Life of a Cockchafer.*—*From Blanchard.*

g, Young grub feeding in the earth ; *c*, cocoon; *c'*, a cocoon cut open to show pupa of the cockchafer with beginning of wings. Above ground the cockchafer is shown both walking and flying. *e*, Elytra, or wing-covers.

eggs, which, at the end of about five weeks, were hatched and became blind white grubs (*g*, Fig. 86) with six slender black legs and hard horny jaws. After a short time these grubs set to work to gnaw

* *Melolontha vulgaris.*

INSECT SIPPERS AND GNAWERS. 253

the tender roots of the young summer plants, and during the next three years fed vigorously underground, eating first what was near them and then making galleries in all directions, and devouring the roots of strawberry plants or rose-trees, oats or corn, or clover, till many were devoured themselves by moles and hedgehogs, or, if they ventured too near the surface of the ground, by rooks, crows, and magpies, which sit upon the clods and pick them out of the loose ground. Those which escaped—and they are usually many—burrowed down deep in the winter out of the way of frost and wet, to come up again in the spring to feed afresh. But at the end of the third year, after having shed their skins several times, they laid themselves down to rest in the earth, and giving out a kind of sticky froth, which they bound with threads of silk into a cocoon (*c*, Fig. 86), they split their last grub skin and remained as pupæ or swaddled insects (*c'*), with their imperfect wings folded over their legs and antennæ. Then early in the fourth year, about April, the true cockchafer began to stir in the cocoon and crept out of the ground, hungry with its long fast, and flying up to the trees began to gnaw and eat for the short two months remaining of its life, and it is then that we meet with it flying from tree to tree, and browsing with its strong mandibles on the leaves of the oaks and beeches and maples.

The history of the cockchafer is that also of many other beetles. The grubs of the beautiful golden green rose-beetle, and many others, live underground, feeding on the roots of plants, and the great stag-beetle whose sharp jaws as a grub enable him to eat into solid wood, only makes this difference, that he spends his three or sometimes even six years of childhood in the trunk of an old oak-tree, gnawing away at it for his daily meal, and only sees daylight when he eats his way out as the perfect beetle.

Fig. 87.

The Nut Weevil.*

w, The perfect weevil; *m*, head of the maggot eating its way out of the nut.

But the little weevils with their curious snouts (Fig. 87), which they use for piercing holes in which to place their eggs, love best the centre of flowers or tender leaves, or especially fruits and nuts of various kinds, for their nursery. When we crack a nut, and find a fat white maggot inside, we have disturbed the forerunner of one of these little weevils, which,

* *Balaninus nucum.*

if the nut had remained on the tree, would by and by have worked its way out (*m*, Fig. 87), and fallen to the ground, where it would have gone to sleep all through the winter, to wake with a long thin snout, and a pair of delicate wings hidden under its beautiful brown wing-cases. The pea-maggot, on the other hand, would, if we had left it alone, have lain down just within the delicate skin of the pea, and there been transformed into a tiny brown beetle spotted with white.

Many of the weevils do indeed eat the bark, and wood, and roots of trees, for they are a very numerous family, and must find food where they can, but the greater number of them feed on fruits, buds, flowers, and grains of all kinds, so that you need only hunt among the acorns, and wheat, and rape, and turnips, to make acquaintance with these tiny beetles; or if you seek out the faded dingy-brown blossoms on an apple-tree, which remain when the other bright blossoms are turning into fruit, there you may find either a tiny chrysalis, or a short-snouted weevil, which has lived all its life in this blossom since its mother laid the egg in the early spring, and whose food, as a maggot, has been the tender centre of the flower.

These are all plant-eating beetles, and they, or some of their comrades, may be found on every plant or tree, nay, you may even shake a shower of them out of the folds of a large mushroom, though they are so small you must get a microscope to see them.

But the Tiger-beetle with its brilliant golden green wing-cases, the Bombardier-beetle (see Plate II. p. 135) which shoots out a vapour from its tail when it is attacked, the common garden beetle or Carabus which pours a black fluid on your fingers when you catch it, and even the delicate little Ladybird, which is a true beetle, are all animal feeders, and they destroy a whole host of insects, such as aphides, caterpillars, weevils, cockchafers, centipedes, and flies. The young wingless lady-bird creeps after the aphides, eating them one by one up the stem (as we saw the blind-grub of the fly doing in Fig. 69), while the grubs of the tiger-beetle have a most cunning way of catching their food, for they bury themselves in the soil with their mouths just above the ground, so that the ants and small insects run heedlessly into their jaws.

These, and many other beetles, feed greedily upon living creatures, and are quite as eager hunters of small animals as lions and tigers are among large ones. You need only watch the ugly cocktail beetle (Plate II. p. 135) scampering after some insect, or seizing upon one of its weaker brethren as it cocks up its head and tail, and snaps its sharp jaws, to understand how aggressive these creatures can be. Among the water-beetles too, though some, such as the black water-beetle,* are vegetarians, yet many are most voracious and cruel; the true water-beetle† (Fig. 88) which dives and swims so powerfully with its broad hind legs, and carries air under its closed wing-cases, is one of the most greedy of water animals, both as a grub and beetle. Not only does it devour the grubs of may-flies, dragon-flies, and other pond insects, but it feeds on snails, tadpoles, and fish; taking care, however, to burrow deep in the earth out of

---

* Hydrophilus. † *Dyticus marginalis.*

the way, when its helpless season comes, lest some of these creatures should return the compliment.

This beetle is strong and powerful, and looks like a dangerous enemy; but who would think that the

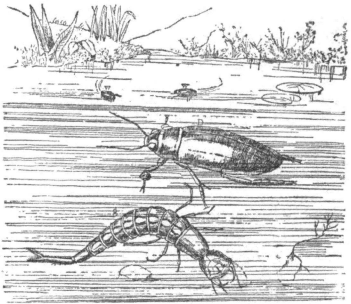

Carnivorous Beetles. Natural size.

D, True water-beetle;* g, grub of the same, showing its powerful pincers and strong head; w w, whirligig beetles.†

tiny bronze-coloured whirligig beetles† which look so bright as they dance on the top of the pond, are also voracious insect-feeders! Watch a group of these bewildering little animals carefully, and you will see one here and one there dart up to catch

* *Dyticus marginalis.* † *Gyrinidæ.*

a passing fly, or down rapidly to scize some tiny water insect, or to escape an enemy that is approaching. For, in fact, these beetles have an unfair advantage in life, having each of their eyes divided in two parts, one half looking down into the pond below, and the other half up into the air, so that they can literally keep "half an eye" upon any suspicious creature in either element. Theirs is a life of many experiences, for after beginning their existence on the surface of water-plants, where the mother places the eggs, they dive down as grubs to the bottom of the pond, breathing by hairy gills, and leaping actively here and there by four curious little hooks on their tails, feeding vigorously all the while; then they creep up into the air, and spinning fine cocoons upon the leaves of a water-plant, remodel their bodies; and finally, as tiny beetles they lead a giddy life on the pond-surface, darting here and there as fancy guides them.

But, quick as these and many other water and land beetles are, both in catching and escaping other animals, it is a curious fact that it is among the scavenger and filth-feeding beetles that we must look for the highest intelligence these creatures possess. It is the dung-feeding beetle, the sacred Scarabæus of the Egyptians, which rolls up a ball of dung with her hind legs, and then sometimes alone, sometimes in company with another beetle which hopes to share or steal the booty, rolls the ball to a convenient place, and digging a hole by means of the large spines on her front legs, buries it and herself with it, so that she may feed upon it in safety. Then later on in the year she hollows out a closed chamber and fills it with prepared dung in which to lay her egg.* It is again among the carrion beetles that we find the "Sexton," burying dead animals carefully under the soil, and then laying her eggs in them.

The history of these sexton beetles is most extraordinary. They hunt in couples, male and female, often many couples together, and wherever they find a dead bird or mouse, rat or frog, they first feed till they are satisfied, and then drag the body to a soft place in the ground. Here the male beetles set to work, and with their strong heads dig a furrow all round the animal, then another and another, till little by little the carcass sinks down, so that actually in about twenty-four hours it is below the ground, and they can cover it with earth, burying the mother beetles with it. Then the fathers too burrow down, and all is quiet and still—but not for long, for no sooner has the mother beetle laid her eggs in the dead body safe out of sight of all enemies, than both mother and father make their way out of the earth and fly away. Meanwhile, the eggs left in the decayed body are soon hatched, and the grubs feed for three or four weeks, and then each building a cell, lies down to undergo its change, and comes out of the earth a perfect sexton beetle.

At first sight it seems almost impossible that such small creatures can bury others so much larger than themselves, yet Miss Staveley,† a good authority, states that four of these beetles have been known to bury in fifty days, four frogs, three small birds, two

---

* For an interesting account of these beetles, showing that the idea of an egg being contained in the rolling ball is erroneous, see M. T. H. Fabre, *Souvenirs Entomologiques*, Paris, 1879.

† *British Insects*, p. 74.

fishes, one mole, two grasshoppers, the entrails of a fish, and two pieces of ox-liver! Which among us works harder than this to provide food for the little ones who come after us? Or who can say that these little beetles do not do their share of good in the world, when they clear away masses of decaying matter which would poison the air, burying it in the best of all purifiers, our mother earth?

So feeding on plant or animal, in the land or in the water, the beetles, with their strong-jointed legs and powerful jaws, make their way in life. You have only to watch a beetle forcing its way under a clod of earth, to see how powerful their muscles are; indeed, it has been estimated that a cockchafer can draw a weight fourteen times as heavy as itself, while the bee-beetle* can draw forty times its own weight, and many of the feats of beetle-life beat those of any athlete among men. Yet we find that they are not wanting in cunning too, for who has not seen the common skip-jack beetle drop on the ground when alarmed, and drawing in its legs and antennæ, lie on its back, and pretend to be dead till the danger is past, and then with a sudden click of its breast-plate, spring up in the air and come down upon its legs? But we must pass by many of these curious histories, such as that of the parasitic beetles which introduce the eggs of their young into the bee's nest, where they feed upon the honey, and of the blind beetles which live among the ants, and must even neglect the soft-skinned glow-worms with their phosphorescent light in the last three rings of the abdomen, and the beautiful fireflies of warm

* *Trichius fasciatus.*

tropical countries, which are near relations of the skip-jack, and have two bright shining spots upon their shoulders. We might trace out in the lives of many of these beetle families the peculiar shape of jaws, legs, antennæ, and the peculiar colours of their wing-cases which fit them for the work they have to do, but such knowledge is the work of a lifetime, and at least a few words at the end of this chapter must be given to the third group of animals which remodel their bodies, namely the two-winged* flies and gnats.

Does it not seem strange that while butterflies and beetles, dragonflies and grasshoppers, and even bees and wasps, have all two pairs of wings, yet our common house-fly and bluebottle, in many other ways so like bees, have only one pair? This, however, will not seem quite so strange if you look carefully just behind the wings of the fly, for there you will find on each side a little stalk with a knob at the end, which the creature uses to balance itself as it flies. These two stalks are the remains of the second pair of wings, which, for some reason unknown to us, must have been a disadvantage to the ancestors of the fly, and this is all that remains of them. If you cannot find them easily in the fly, where they are concealed under some little winglets, you will see them clearly in a gnat, or, better still, in a daddy-long-legs, in which they are so distinct that you may examine them without catching or hurting him, by simply putting a tumbler over him where he stands and slipping a piece of paper underneath.

* Diptera (from *dis*, twice; *pteron*. wing).

These "balancers" tell us that the two-winged flies, the gnats, mosquitoes, midges, bluebottles, house-flies, and cattle-flies, are not made on a different plan from the four-winged insects, but are merely flies whose hind wings have lost their size and power, while the front ones have become stronger and larger. This has evidently been no disadvantage in their

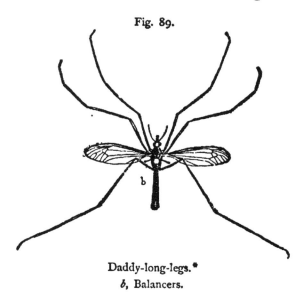

Fig. 89.

Daddy-long-legs.*
b, Balancers.

case, for they have flourished well in the world, and myriads are to be found in every town and country, while their different ways of living are almost as various as there are kinds of fly. Some, such as the daddy-long-legs, suck the juices of plants, some suck animal blood, some live on decaying matter; while in not a few cases, as among the gadflies, the

* Tipula.

father is a peaceable sucker of honey while the mother is bloodthirsty.

Among the gnats and mosquitoes the father dies so soon that he does not feed at all, while the mother has a mouth made of sharp lancets, with which she pierces the skin of her victim and then sucks up the juices through the lips. Among the botflies, however, which are so much dreaded by horses and cattle, it is not with the mouth in feeding that the wound is made. In this case the mother has a scaly pointed instrument in the tail,* which she thrusts into the flesh of the animal so as to lay her eggs beneath its skin, where the young grub feeds and undergoes its change into a fly.

For we must remember that every fly we see has had its young maggot life and its time of rest. Our common house-fly was hatched in a dust heap or a dung heap, or among decaying vegetables, and fed in early life on far less tasty food than it finds in our houses. The bluebottle was hatched in a piece of meat, and fed there as a grub; and the gadfly began its life inside a horse, its careful mother having placed her eggs on some part of the horse's body which he was sure to lick and so to carry the young grub to its natural warm home.

But of all early lives that of the gnat is probably the most romantic, and certainly more pleasant than those of most flies. When the mother is ready to lay her eggs she flies to the nearest quiet water, and there, collecting the eggs together with her long hind legs, glues them into a little boat-shaped mass and

* A similar instrument may be seen in the daddy-long-legs if you happen to catch a female; she uses it to thrust her eggs into the earth.

leaves them to float. In a very short time the eggs are hatched and the young grubs swim briskly about, whirling round some tufts of hair which grow on their mouths, and so driving microscopic animals and plants down their throats. Curiously enough they

Fig. 90.

Life of a Gnat.

g, Grub breathing air through the tube *t*; *p*, pupa breathing air through two tubes *t* in the back; *b*, floating boat formed of the pupa skin; *gn*, gnat rising out of it; above the perfect gnat is on the wing. These figures are all magnified to give clearness.

all swim head downwards and tail upwards (*g*, Fig. 90), and the secret of this is that they are air-breathing animals and have a small tube at the end of their tail, which they thrust above water to take in air. This goes on for about a fortnight, when, after they have changed their skins three times, they are ready to remodel their bodies. Then on casting their skin for the fourth time they come out shorter and bent and swathed up, but still able to swim about though not to eat. Meanwhile a most curious change has taken place. The tail tube has gone, and two little tubes (*p t*, Fig. 90) have grown on the top of the back, and through them the tiny pupa now draws in its breath as it wanders along. At last the time comes for the gnat to come forth, and the pupa stretches itself out near the top of the water, with its shoulders a little raised out of it. Then the skin begins to split, and the true head of the gnat appears and gradually rises, drawing up the body out of its case. This is a moment of extreme danger, for if the boat-like skin were to tip over it would carry the gnat with it,— and in this way hundreds are drowned—but if the gnat can draw out its legs in safety the danger is over. Leaning down to the water he rests his tiny feet upon it, unfolds and dries his beautiful scale-covered wings, and flies away in safety.

With the gnat we must take our leave of the two-winged flies, although if we could study their whole history we should find them so intelligent that we should not be surprised at Mr. Lowne's statement that, although a fly is not one-fourth the size of a beetle, its brain is thirty times larger. In fact it is among these creatures which undergo metamorphosis that we begin to reach a point of intelligence which, of its kind, is quite as remarkable as that of the back-boned animals. But it is not among the butterflies, beetles, or even the two-winged flies, that the highest instincts are found. There exists an immense order called the Hymenoptera, or membrane-winged insects (*hymen*, membrane; *pteron*, wing), including the gall-flies, saw-flies, ichneumons, burrowers, bees, wasps, and ants, in which instinct and intelligence exists to such a great degree that all naturalists are lost in wonder at the ingenuity of the wasp or the bee, and the almost incredible sagacity of the ant.

And here comes a curious fact which we find equally among the insects and the back-boned animals. As Life endows her children with more intelligence, with quicker brains governing active bodies, we find them becoming more and more dependent upon others in their infancy and youth. Just as the large and man-like orang-outang remains as helpless as a human baby for the first few months of its life, while the lower and less intelligent monkeys have, long before that age, begun to fight their own battles;[*] so while the grubs of the frivolous butterfly, the thoughtless gnat, and even the more intelligent saw-fly, are active and can take care of themselves from the time they come out of the egg, the cell-building bee and wasp on the contrary, and the thoughtful contriving ant, have a real babyhood, during which others watch and tend them, and when they must perish, just as a child would, if it were not for the care and attention of their grown-up friends. And this helplessness of infancy increases with the intelligence of the grown-up creature, as we shall see on reading the next chapter. For no one will deny that the ant stands first in mental capacity among insects, and its

[*] For an amusing account of the difference between an orang-outang baby and a young harelip monkey of about the same age, see Wallace's *Malay Archipelago*, p. 45.

children are more helpless even than those of bees. A young bee eats its own food placed for it in its cell, but the ant can take nothing but what is actually put into its mouth.

It is most tempting to try and trace out this gradual progress to increased intelligence in age and helplessness in youth among the membrane-winged insects. Thus we should begin with the caterpillars of the saw-flies, placed within their proper plant by the saw-like instrument of their mother, and creeping over it in their youth; then pass on to the grubs of the gall-flies which lie helplessly within the gall-nuts eating the food which their mother has prepared for them by leaving an irritating liquid which causes a lump to grow up around them on the plant. Next would come the grubs of the cunning Ichnuemon fly which, though feeding on honey herself, pierces the skin of the caterpillar or the beetle, and leaves her eggs in their flesh, where the young ones live as parasites during their sluggish infancy.

From these we should go on to the still more wonderful burrowing insects, such as the Cerceris, the Sphex, and the Sand-wasp, which, after laying their eggs in a hole, pierce beetles, grasshoppers, or caterpillars with their sting, not killing them, but paralysing them, and then storing them up with their eggs as fresh healthy living food for the young when they are hatched, two or three weeks later. Then we should come to the true wasps, with their beautifully-constructed paper nests, built of wood fibre moulded into paste, and their helpless infants each in its cell tended with the utmost care; and we should learn almost to have an affection for these industrious creatures, which in some ways show even greater intelligence than the bees. Then these last would claim our attention, with their frugal habits, their industry in storing up honey, their wonderful cities, in which each citizen has his duty, and their love for their queen. And, lastly, we arrive at the ants, and to these we must devote the next chapter, since to speak of the others would need a whole book, and the bees we have dealt with elsewhere.* In the ants we shall find that life has worked out her masterpiece among insects, and in them we can best learn to understand how far we have travelled, since we started with the Amœba, passing gradually from mere living, feeding, and dying atoms of life, to active, intelligent beings, whose life depends quite as much, and even more, upon the inward work of the brain than upon the outer weapons of the body.

* *Fairyland of Science.*

CHAPTER XII.*

INTELLIGENT INSECTS WITH HELPLESS CHILDREN AS ILLUSTRATED BY THE ANTS.

> "So when the emmets, an industrious train,
> Embodied, rob some golden heap of grain,
> Studious, ere stormy winter frowns, to lay
> Safe in their darksome cells the treasured prey,
> In one long track the dusky legions lead
> Their prize in triumph through the verdant mead,—
> Here, bending with the load, a panting throng,
> With force conjoined, heave some huge grain along,
> Some lash the stragglers to the task assigned,
> Some to their ranks the bands that lag behind;
> They crowd the peopled path in thick array,
> Glow at the work, and darken all the way."
>
> VIRGIL.

"I DARE engage," said the King of Brobdingnag, as he took Gulliver on the palm of his right hand and stroked him gently, while his learned men examined this strange pigmy through their magnifying glasses, "I dare engage that these diminutive creatures have their titles and distinctions of honour; they contrive little nests and burrows which they call houses and cities; they make a figure and dress in equipage; they love, they fight, they dispute, they cheat, they betray."

---

* Most of the facts in this chapter which are not to be found in the standard works of Huber and Gould, have been taken from the works of Forel, McCook, Belt, and Moggridge, and from the scientific papers published by Sir J. Lubbock.

Now Gulliver was a man made in the same fashion as the Brobdingnagian king, only in smaller proportions, and therefore it was not so wonderful that the king should suppose him to be living a life like his own. But the little ant, which we may take in like manner on our hand, is fashioned quite differently from ourselves, and is only an insect; and yet, strange to say, almost all the things asserted of Gulliver, and many more which seem almost human, may be said with truth about these tiny creatures.

For ants make nests and burrows which are real houses and cities, and even clear roadways to and from their settlements. If they do not dress in equipage, they perform their toilet with the greatest care, as well as that of their friends and of the helpless infants of their city. They can dispute and hold communication with their fellows; they fight, both singly and in well-disciplined armies; they betray in some cases their fellow-ants, and carry them into slavery; they keep domestic animals, having beetles and other insects living in their nests, as we have dogs and cats in our houses, and some of them provide food for their community by keeping herds of Aphides as we keep cows. Moreover, they form very large societies, such as can only succeed by all the members working together in harmony. In one ants' nest, made up of several separate homes, there may be from 50,000 to 200,000 ants, and though each ant is free to build, or hunt, or milk, or fight, or go where she will, it seems to create no confusion. In this way they are much more independent than bees and wasps, whose combs are all built mechanically, one exactly like the other; whereas any ant

ANTS AND THEIR HELPLESS CHILDREN. 271

may start a gallery or chamber in a new direction, and others soon joining her will add fresh nurseries and hom s to the nest according to their own ideas. Yet there is order in this vast multitude. Some invisible bond makes each and all labour for the good of the whole, and this is the more curious as

Fig. 91.

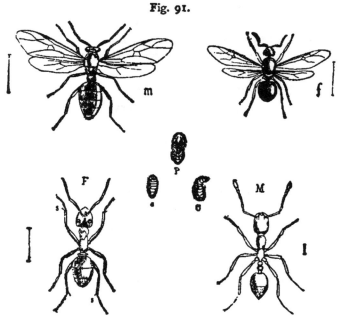

F, *Formica rufa.* Hill Ant. Worker. *s*, Spur; *m*, male; *f*, female; *g*, grub; *c*, cocoon; *p*, naked pupa.
　　M, *Myrmica molesta.* The little reddish-yellow Ant, infesting our houses, having two knobs and a sting.
　　The true size of these Ants is indicated by the lines.

there is no special leader or governor among them. In all complicated work which has to be done—the feeding of the queen-mother, the nursing, carrying, and feeding of the young, the building of new chambers, the tending of their flocks and herds, the defence of the nest, or the formation of new colonies —all labour amicably together, without any apparent government, and yet without confusion or disorder.

Now before we can understand how it is that these little insects have advanced so far beyond others of their class, we must first inquire what are the chief weapons with which life has provided them, and to what use they are put. Although ants are such common insects that there is scarcely a garden without them, and even many of our houses are overrun by them, yet probably very few people have ever examined one carefully, or tried to understand its very peculiar shape; and if you can catch one wandering about the garden or feeding in the store-cupboard, and put it under a magnifying glass, you will be astonished to find how much there is to learn about it.

First notice the ringed abdomen common to all insects, and the very fine stalk by which it is joined to the rest of the body, allowing it to bend easily in all directions. If you have taken the garden-ant,* this stalk will be made of one knob or ring, as it is also in the hill-ant (F, Fig. 91), while, if you have the tiny reddish-yellow house-ant† (M), there will be two knobs; and by this you may know at once that the house-ant has a sting, while the garden one has not; for in England all the ants with one knob, except one single genus,‡ have no sting. Next notice the three-ringed thorax bearing the six legs. On each side of this are three breathing holes, which,

* *Lasius niger*. † *Myrmica molesta*. ‡ Ponera.

ANTS AND THEIR HELPLESS CHILDREN. 273

however, you cannot see without a very strong lens. But even with the naked eye you may discover the tiny spur (*s*, F, Fig. 91), which sticks out from the third joint of each leg, and a magnifying glass will show that this spur on the two front legs is larger than on the others, and bears on its edge more than fifty-five elastic teeth (*c*, Fig. 92), while another set of similar teeth on the leg itself (*lc*) face it, and can be rubbed up against it. These are the toilet-brush and comb of the ant; and whenever she has been doing any dirty work, she will pause, and use them to brush off the dust or mud which has clung to the delicate hairs and bristles of her body. Then she will afterwards pass the brush and comb through her mandibles, and so clean them afresh for work.

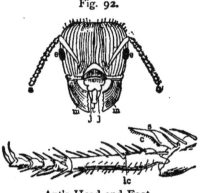

Fig. 92.

Ant's Head and Foot.

Head—*e e*, eyes; *a a*, antennæ; *m m*, mandibles; *j j*, jaws; *t*, tongue.

Foot—*s*, spur; *c*, comb of spur; *l c*, legcomb.

It is, however, the head of the ant which is above all remarkable. You will be struck at once with its curious triangular shape, its large size, and its flatness, while the small eyes (*e*, Fig. 92), and the antennæ (*a*), bent like the elbow of an arm, are very different from what we have seen in other insects. It seems strange at first that active intelligent creatures like ants should have such small eyes as many of them have, and still more so that the eyes of the workers should be smaller than those of the males and females which do no work. But when we remember the blind Termites (p. 225), and how they build intricate homes without any eyes at all, we are prepared to find that it is the antennæ which are chiefly used by ants to guide them in their work.

What the true history of these antennæ is, and how the ants communicate by means of them, we shall probably never know; for though it is almost certain that they use them for feeling and smelling, and perhaps even for hearing, yet there seems to be some other sense in them by which one ant can tell another of danger, or food, or work to be done.

For instance, Sir J. Lubbock, who is unwearying in making careful and accurate observations on the habits of ants, has lately tried the experiment of pinning a fly or a spider to the ground, so that the ant which found it could not drag it away. On nearly all occasions the ant returning to the nest brought friends to help her; seven, twelve, and in one case fifteen came; and as she did not carry anything with her to *show* that she had found a prize, it is almost certain that she must have told them of it in some way. The ants which she brought often came slowly and reluctantly, wandering hither and thither so as to be half-an-hour in reaching the dead insect, and once the first ant growing impatient started off again to the nest and brought a second body of recruits, who "after most persevering efforts carried off the spider piecemeal." These ants, then, had some means of telling the other ants that they wanted help, and how much they wanted; and numerous observations show that it is by touching their antennæ that they make these communications. Now, though at first this may appear almost incredible, yet, if we think for a moment, we shall acknowledge that it would seem still more strange to a being who knew nothing about speech, to see two men stand at a distance from each other, and only move their mouths, and then go and do something which showed that one knew what the other wanted, so that it may after all be only our ignorance of ant-language which puzzles us.

Next to the antennæ, the most useful implements of an ant are her mandibles (*m m*), which do the greater part of the work, to which the antennæ guide them. Looking in the face of an ant, you see these two outer jaws, with their toothed edges resting against each other; but if you make her angry, she will open them wide to seize you with all her tiny might.

Does she want to excavate a gallery? Then she will tear out the earth with these toothed spades, and carry it in pellets above ground. Is she cleaning a cocoon? She will then use her mandibles, tenderly and neatly, to pick out morsels of dirt, and afterwards will lift the tiny ball with them, and carry it without injury up or down the nest. Or she may perhaps be cutting a blade of grass to lay as a rafter in the roof of a chamber; again she saws off the leaf with her mandibles, while she holds it with her front legs. Or, lastly, if she is fighting to the death in a pitched battle, she will fix these strong pincers so firmly in the throat of her enemy, that even if she is killed, her head will often remain for days hanging on to the conqueror's body.

There is, however, one thing she does not do with her mandibles; she does not chew with them, but uses them to tear and press the food so as to obtain the juices and oils in it. It is true many ants feed upon other insects, and even on grains, but in the first case they pierce the skin with the mandibles, and then lap up the liquid within, and the seeds they tear to fragments, and lick or rasp off the starch with their tiny tongue (t, Fig. 92), helped by the inner jaws (jj).

Such, then, is roughly the structure of the working ant, which is an imperfect female; and when we ask how it is that so small a creature, with a body not one-tenth part as strong as many of the beetles, and without the power of flying, has made its way so well in the world, we learn that within that curious-shaped head is collected a larger and more complex mass of nerve-matter than in other insects, so that in the two large hemispheres of an ant's brain, life has prepared a powerful machine for guiding the little creature on its road. In all social insects, such as the bees and wasps, the nerve-masses in the brain are larger than in those insects which do their work alone, and one great secret of the success of ants is that they form the most perfect societies in the whole animal world.

And now, how shall we study ant-life? For there are as many different races of ants, each with its special habits and customs, as there are races of men, and one description will by no means do for them all. The best way will be to speak first of some one race well known to all of us, and then to say something of others. It would seem most natural to take the little garden-ant, which is the one we most often come across. But it lives a great part of its time underground, and though it comes to the surface to sun itself and wander about, it does not do much work above ground, except when it is visiting its cows (see p. 287). It will be better therefore to take another common ant, the hill-ant or horse-ant,* as it is often called, which lives a more out-door life.

You can scarcely walk through any English wood without coming across lines of these reddish-looking ants, which are often of very different sizes, and have, for ants, rather large eyes. Their nests are easily found, forming large leafy hillocks at the foot of oak trees, or sometimes in the open ground. Even in England they are often two feet high, and on the Continent they are much larger. At an early hour in the morning all will be still and quiet on these hillocks, for the ants close their doors at night with leaves, or bits of stick and straw: but as soon as the sun rises and flings its beams across the leafy wood, warming the air, you may see a few ants creep out of cracks in the dome; and by and by, if the day be fine, many large openings will be made, and soon all is alive and active. Some ants are dragging in bits of wood, and straw, and leaves, to add to the dome; others are carrying in bits of insects, young grasshoppers, or worms, or caterpillars, whose juices they will feed upon in the nest; others creep into the blossoms of plants to steal their honey; while others, again, seek out the stems covered with aphides or plant-lice, and beg of them their sweet juice.

* *Formica rufa.*

It will be remembered (see p. 203) that the aphides plunge their trunks into plants, and suck all day long, filling their bodies with juice. Now, when the ant comes running up the stem in search of food, she comes behind the aphis and strokes it gently with her antennæ, and the little creature gives out from the end of its body (or sometimes from the little horns), a drop of sweet liquid which the ant licks up, and it is probable that this is pleasant to the aphis, which in any case always gives out juice from time to time. The ant, on her side, protects these plant-lice, keeping off the lady-birds or other insects, which might attack them, and even taking care that, for a certain distance round her own nest no ant from a strange community shall poach upon her grounds.

And now, as these well-fed ants, with their crops filled with two or three drops of aphis juice, hurry home again they meet with others, those that have been collecting leaves, or those which have been sweeping out the galleries of the nest, and have had no time to get food. These hungry ants run up to the full-fed ones, and stroke them with their antennæ, asking for food, and then lifting up their mouth, they receive the juice which the others squeeze out of their crop; for one of the principal rules in ant-cities is for every member to help another for the general good.

Busy, however, as every one seems to be outside the nest, they are still busier within. If you could cut one of these ant-hills in half downwards, you would find that the nest extends often a foot or more into the earth, and everywhere it is a maze of narrow

ANTS AND THEIR HELPLESS CHILDREN. 279

galleries leading into rooms with vaulted ceilings (see Fig. 93). In the top part of the nest, made chiefly of sticks and leaves, together with dried seeds and often little stones and shells, these galleries appear very confused, though with a little care they may be traced by the tiny beams of wood, and the blades of grass and leaves forming the rafters of the

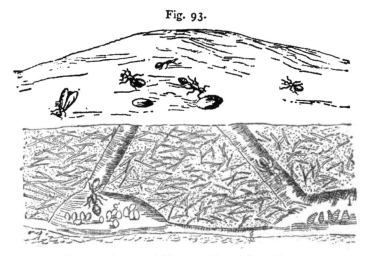

Fig. 93.

Section of an Ants' Nest.—*Adapted from Figuier.*

g, Gallery; *c*, cocoons in a vaulted chamber; *l*, larvæ or young ant-grubs.

ceilings; but, down below, where the ground is firm though still mixed with other material, the roads are clearer and the chambers larger.

Here active busy work is going on. Deep down, almost at the bottom of the nest, the queen-ant is wandering about with her train of followers, dropping tiny eggs as she goes, which the workers pick up and arrange in little heaps in the chambers. In other apartments are packets of eggs, many days old, and these are being licked all over and carried, several at a time, by the workers up into higher chambers, where the air is warmed by the morning sun. Again, in other chambers are heaps of little white, legless, blind grubs, with twelve soft rings to their bodies (g, Fig. 91 ; l, Fig. 93), and narrow mouths with soft mouth-pieces ending in a pointed lip. These little helpless creatures can do no more than just turn their heads to receive the drops of food which the nurses squeeze out of their crops down the infant throats. They are spotlessly clean, for they too are licked all over daily, and every speck of dirt is picked off by the mandibles of the worker ants, which not only feed and clean them, but carry them as they did the eggs, up for warmth in the day, and down at night to escape the chilly air. Sir John Lubbock has observed that these grubs are sometimes even sorted and arranged in groups, according to their size and age ; for they live and grow in this state for various periods according to the time of year, and sometimes remain as grubs for many months.

In another chamber, quite a different process is going on, for here the grubs have arrived at the time when they are ready to remodel their bodies ; and each little grub, moving his head to and fro, is laying down silken threads within which he spins his soft cocoon. Still, here also the workers are busy, for as soon as each cocoon is finished, they loosen the outer threads clinging to the earth, and smooth and clean the cocoon till it is a pure oval ball, which they can carry up and down in the nest, though they can no longer feed the little creature within. It is these cocoons (*c*, Fig. 93) which people mistake for eggs, when they see the ants hurrying away with them when their nest is disturbed; for the nurses guard their sleeping children with zealous care, and many a worker-ant has died sooner than leave a grub or a cocoon in the hands of an enemy.

Lastly, in other chambers the final act of the baby ant's history is being carried on; for after cleaning and carrying and watching over the cocoons till the perfect ant is ready to come forth, the workers have still to help it out of its silken prison. This they do by tearing the cocoon gently with their mandibles, two or three of them at a time. Then carefully drawing the ant out of the hole, and licking it all over to clear it of its pupa skin, they feed it and leave it to go to its work, which for some little time will be all within the nest, till its coat has become hard and firm, and its limbs strong.* When once it is grown up it may live through many seasons; for Sir J. Lubbock tells me that he has ants which have lived in his room since 1874, and they must therefore be now at least six, and probably seven years old.

All this different work of nursing and feeding may be going on at one time in a nest; sometimes in different chambers, sometimes pell-mell, eggs, grubs, cocoons, and young ants all in the same room. But this is not all which the workers have to do, for if it be summer time a number of

* Among some ants the grubs do not spin cocoons, but remain naked pupæ like the chrysalis of the butterfly (see *p*, Fig. 91, p. 271), and even among these hill-ants this will sometimes happen late in the year. When the pupa is naked the young ant can get out by itself without help.

winged ants will be wandering about which have also to be cared for. These are the male ants, and the young females which have not yet begun to lay eggs. They have come out of cocoons rather different in size and shape from those of the wingless worker ants, and we do not yet know what decides this difference. There is no jealousy in an ants' nest (as there is in a beehive) between the queen-mother and the young princesses; indeed in some nests several queen mothers live amicably together. But still the workers have to feed and watch all these winged ants, and though the young princesses are allowed to go outside on the dome of the nest and sun themselves, the workers never leave them, and towards evening may be seen taking hold of them by the mandibles and dragging them gently home to bed.

By and by, later in the year, all these winged ants will come out of the nest in a swarm and rise and fall in the air like the May-flies over the pond. Then the males will never return to the nest, but will wander about and soon die, or be devoured in numbers by birds or other insects. The same thing will happen to many of the princesses, but some will be seized by the workers and dragged back to the nest, where their wings are pulled off and they settle down into queens, and lay eggs. Others which have fallen at a distance will pull off their own wings, which are fastened very lightly to their shoulders, and will begin to dig a hole and lay eggs in the earth. Whether these solitary queens are able to found a new nest, or whether it is only when two or three workers join them that they live and flourish, is not yet certain, but Sir John Lubbock has shown that in one case at any rate a queen which he kept artificially (*Myrmica ruginodis*) did bring up some young workers from her eggs.[*]

There is another way, however, in which new nests begin, and this is when an old nest is over full, or when the leaves and sticks begin to decay, and the carcasses of insects and of dead ants which have been thrown out of the nest make the home unhealthy. In either of these cases some worker sets off and finds a spot for a new nest; this found she comes back, and seizing another ant in her mandibles carries her off to the chosen place. These two again return, each carrying another, and so a little band of workers is collected.

Then they set to work. One ant begins to dig a hole with her front legs, throwing out the dirt behind much as a dog does; another and another follow in her train, and the work goes on merrily, while others are still going to and fro to the old nest and carrying in new recruits. As soon as the tunnel becomes too deep for the earth to be kicked out, the excavators bite out pieces with their mandibles and carry them in little pellets outside the hole to form the upper part of the nest. Meanwhile others are fetching sticks and leaves to prop up the galleries and roof the chambers; and so the dome rises above, as the firmer part of the nest is being scooped out below.

And now the migration goes on apace; no ant seems to leave the old nest willingly, but as soon as she is carried to the new one the general enthusiasm seizes upon her, and she sets to work to dig and build, or runs back to carry another as eagerly as

[*] "Habits of Ants," *Science Lectures*, p. 92.

any of those that have been long at work. In this way a complete train of ants going and coming are to be seen, those which are empty-mouthed going back to the old nest, the others each with her burden going to the new one. M. Forel once counted from forty to fifty in a minute passing each way, so that 36,000 new-comers would be carried in one day.

At first it is only workers that are brought, but when chambers are prepared, then the eggs, larvæ, and cocoons, princesses, males, and queens, are all carried to the new home, and the migration is complete.

The next step, if the community be large, is to clear roads to the nearest plants where aphides may be found, and to do this the workers carry away dead leaves and refuse matter, and saw off the living blades of grass, and soon lay bare a narrow path along which they travel in search of food. Now, while all this is going on, it will often appear as if any one ant was wasting a great deal of time running hither and thither in an aimless kind of way. But it must be remembered that ants see but a very short distance, and that it is by means of their antennæ, and probably chiefly by scent, that they find their way. Moreover, the blades of grass among which they are moving are to them like tall trees, so that we must look upon them as travellers in pathless woods following a track, and not expect them to go direct to their point. Many too will seem to stand idle, while others look as if they were merely playing together. This is because they not only rest from time to time, but are very particular to clean their bodies carefully from the earth which clings to them, and when they appear to be playing together, one ant is often performing this kind office for another. Still, in spite of wandering and resting and cleaning themselves, it is marvellous what an amount of work these little creatures do, so that in a short time their new domain is adapted for their life.

It may happen, however, that all does not go so smoothly; the new colony may chance to encroach upon the territory of some other ants' nest, and then comes dire disaster; for no two nations can fight more relentlessly for a province or a frontier than these little ants do for their plots of ground. No sooner does one community find that another is taking possession of any part of its domain, or has given offence in some way unintelligible to us, than the workers pour out by thousands, marching close together in battle array, eager for the attack. Meanwhile those belonging to the other side advance to defend themselves, and when the two armies meet they fall upon each other, two by two, taking hold with their mandibles, and raising themselves up on their hind legs, so that they can turn the abdomen under the body. This is in order that they may squirt out from the tail that strong acid called *formic acid*, which acts as a poison, so that often the combatants fall dead locked in each other's arms. Whenever these ants are excited the scent of formic acid is very strong, so that you may smell it in passing the nest.

If one ant succeeds in disabling the other, remaining unhurt herself, she tries to drag her victim off to the nest, there to be killed and devoured. Now, although all these ants are exactly alike, each army knows its own followers, and it is very rarely indeed that a friend is attacked. If this does happen, the onslaught is almost instantly changed to a caress, and the two friends turn against a common enemy. At night each army returns home, but at daybreak the battle begins again, and may rage for many days till either the inhabitants of one nest are destroyed or routed, or bad weather puts an end to the fighting. And when the war is over, the dead and mangled are not left on the field, for these terrible cannibals carry them off to their nests to suck the juices from their bodies.

Such are the battles of the hill-ants, but the mode of attack is very various among the different races. The red ant (see p. 277), for example, is much more wily and given to stratagem, and does not fight in such large masses. Again, there are tiny ants which, when attacked by larger ones, hang on the legs, and jump upon their backs, biting them and tearing them to pieces, while the larger ant tries to strangle them in her mandibles. One particular slave-making ant* has especially pointed mandibles, and she drives them right into the brain of her enemy, throwing her into convulsions and paralysing her. On the other hand, those ants which have stings make use of them in fighting rather than of their mandibles, while, as we have seen, the hill-ant is remarkable for the force with which she can squirt out formic acid over her adversary.

But, in whatever way they are carried on, these ant-battles are fierce and bitter, for ants have very few enemies but those of their own kind, so that

* *Polyergus rufescens.*

they swarm everywhere, and have great struggles for their homes, and their flocks of aphides. Indeed, among ants, as among uncivilised human races, each member is faithful to his tribe and bitterly hostile to any stranger. Sir J. Lubbock has found that after a separation of fifteen months an ant belonging to the nest was recognised and welcomed, while a stranger was hustled and turned away; and what is still more curious, when ant-eggs were taken to another nest, and the young ones hatched there and brought up by strange nurses, yet their own people recognised and received them when they were returned to their home.

The little black ant of our gardens has learnt a cunning way of keeping out of the way of quarrels by hiding her honey-cows. Instead of going out to seek them every day, she carries them home and keeps them close to her nest, where she sometimes surrounds the stem on which they live with a tube of earth, or visits them by covered galleries, or even puts them on the roots of plants, underground in her own home. You may discover this little ant climbing the plants, and tempting the aphides to give out their sweet drops; and by carefully digging up the plants near her nest, you may find the plant-lice clinging to the roots, which run through her galleries and her chambers. Only, it may be well to put them back again to their industrious keepers, and you will probably be rewarded by seeing the ants take them up, and carry them down for safety to a lower part of the nest. The small yellow meadow-ant* keeps her aphides entirely on the roots of plants, fetching and placing

* *Lasius flavus.*

them near to her nest, and tending them with the greatest care, even watching over their little black eggs, so as to secure fresh broods. In this way she rarely needs to come above ground, and has no regular openings to her nest.

Now, when these ants are attacked, they do not come out of their nest to fight; on the contrary, they defend it like a fortress, hiding themselves in the lower galleries, barring the way with pellets of earth, and disputing every inch with the enemy. Indeed, if the struggle becomes very hopeless, they will escape with their cocoons and grubs along their labyrinth of passages, and closing up the road, will raise a new ant-hill at some distance from the first. These ants work at their nests by night, for as they build entirely with earth, they need the damp and dew to moisten the pellets, as they plaster their walls.

The ants we have mentioned, the hill-ant and the meadow-ant, have workers of very different sizes, and there is very little doubt that the larger workers do most of the fighting; indeed in South Europe and America there are in some species special large-headed workers, which are the soldiers of the community. But if you will search carefully in the banks of the fir woods, or in the stumps of the decayed trees of Hampshire, Surrey, or Sussex, you may chance to come across a much more curious sight than mere difference of size; for you will find large red ants [*] and smaller black ones,[†] living in one nest, and working happily together.

Yet these black ants were not born among the red ones; their eggs were laid by their own black

[*] *Formica sanguinea.*   [†] *Formica fusca.*

mother in the nest at home; and they were stolen the summer before, some time between the months of June and August, when they were lying wrapped up in their cocoons, by an army of red ants which attacked the nest in which they lay.

The first alarm was probably given by the appearance of a few scouts wandering round the nest, and as soon as the black ants saw them, there was terrible consternation. Some swarmed out to face the enemy, others rushed to seize the eggs, grubs, and cocoons, to carry them to the other side of the nest for safety, where the princesses followed them, while others blocked up the doors to defend the fortress. Meanwhile, the red army kept growing in numbers, more and more ants crowding round the poor black-ant city, till a semicircle of hundreds of warriors was formed, all standing with their antennæ thrown back, and their mandibles wide open. Then the attack began; the black ants fought bravely, clinging to the legs of their cruel foes, biting them, and striving to drive them off, and to bear their little ones away in safety. In vain; overcome by the strength and number of their assailants, they had to give way, and soon the conquerors were swarming over the dead bodies of the defenders, and carrying off their cocoons.

From this moment, the red ants did not attempt to fight, except with those of the black ones which tried to escape with their young. They hurried past all the others, pushing them aside, and occupied themselves entirely in making their way into the galleries, out of which each red ant came with her stolen treasure in her mouth, and set off with it at once to her own nest. In this way the whole red army travelled to and fro, carrying away cocoon after cocoon, and delivering them over to the care of other black ants in the nest at home, which had been captured in the same way the year before, and had now settled down as nurses of the establishment. These ants took the cocoons, and watched and tended them, and by and by, when they were opened, the black children took to the red-ant nest as their home, and worked with the rest. Except that they are forcibly seized in their infancy, these black ants can scarcely be called slaves, for master and servant live together like equals, only that the black ants generally remain more indoors, while the red ones go out to seek food.

But how have these red ants, which are in many ways some of the cleverest of their kind, learnt to steal young black ones, to help them in their work? Mr. Darwin suggests the answer. It is a common practice with ants to carry the cocoons of their enemies into their nests to eat them, and they tear open the cocoon to feed on the insect within. Now, nothing is more likely than that some of the black-ant cocoons, thus carried in, should be neglected, till the ants within them were perfect, and then, when they came out active and vigorous, they would be well received, as ants born in the nest generally are, and would mix with the red ones, and prove very useful. Is it too much to imagine, that thus by degrees the intelligent red ants should come to understand that it was better to have the help of the black ants than to eat them, and should learn to fetch them in numbers to help them in their work? One thing is certain, that they know their own interest now, for if by chance a female *winged* ant comes out of the stolen cocoons, she is killed at once by her red masters, who know that if she lived and laid eggs, these would be tended by the slaves, and the nest would soon become a black-ant city.

But now see how true it is, even among insects, that those who always look to others to save them trouble, become weak, useless, and contemptible, for though the slave-making ants which we have in England * work with their slaves, there are others living abroad,† which have become so dependent upon their black servants, that they can neither build their nests nor tend their young, nor even feed themselves. It is a mockery to call the neuters of these ants "workers," for they can do no work any more than the males and females, but they are "soldiers," for the one thing they can do is to go in great hordes and fight the black ants, and steal their cocoons. Here their pointed mandibles (which have lost their toothed edge, and are of no use for other work) come into play as deadly weapons in crushing the brains of their enemies (see p. 286), and their warlike expeditions are bold and successful. For the rest they are quite helpless; it is the negro ants which fetch provisions, feed the grubs, take care of the princesses, build the rooms and galleries, and even feed their lazy masters. Huber once took thirty of these red ants with their grubs and cocoons, and put them in a box with a supply of honey, but though the food lay close to them, they made no

* *Formica sanguinea.* † *Polyergus rufescens.*

attempt to eat, and many died. At last, taking pity on them, he put one little negro ant into the box, and at once she set to work, made a chamber in the earth, fed the grubs, attended to the cocoons, and even saved the lives of those few full-grown red ants which remained. Sir J. Lubbock has lately repeated this experiment, keeping one of these red ants alive for months, by putting in a slave for two or three hours a day to clean and feed it.

Even when it is necessary to migrate to a new nest, these red ants will not trouble themselves to walk there, but lie on their backs, and are carried by the faithful blacks, who never seem to lose their temper, or to mind the work which falls upon them. The only time that we ever hear of the blacks being angry, was once when Huber saw the red ants return from a slave-making expedition without any cocoons. This was too much, that the only one thing they ever did for the community should be neglected! The exasperated slaves hustled them and dragged them out of the nest again, but after a few moments relented and allowed them to come home.

And now from ants degraded by indolence, let us turn to those which have become so industrious as even to lay up stores for the future. Our English ants being in a cold climate sleep through the winter deep down in their lower chambers, and in this way have no need of food; though the yellow meadow ants show great forethought, according to Sir J. Lubbock, by carrying aphides' eggs down into their nest early in October, and tending them with the utmost care, so as to secure a crop of young ones, which they bring out the next March, and place on the daisy stalks, which are their natural home. In warm countries, however, such as the shores of the Mediterranean, and the Southern States of North America, there is no chilling influence to make the ants hibernate, while at the same time for some part of the year they cannot find their accustomed food in the fields and meadows. And so in these warm countries, it comes to pass that the ant " provideth her meat in the summer, and gathereth her food in the harvest," although many writers have denied the truth of this, because they had only studied the ants of colder countries.

It was among the lemon terraces of the warm sheltered valley of Mentone, that Mr. Traherne Moggridge, during the last years of an invalid life, set himself to watch these little harvesters,* which all belong to the two-knobbed ants (see p. 271). There they were to be found in the early spring, as soon as any seeds were ripe, hard at work on the rough slope of the terrace, tearing off the seed-vessels of the Shepherd's Purse and the Chickweed, and plundering Pea-flowers, Honey-worts, and grasses of their seeds, and then carrying them in their mandibles to the nest. Sometimes a young and foolish ant brought in rubbish, and not a few were deceived by some small white beads strewn by Mr. Moggridge on the path. But these no sooner reached the nest than they were hustled out by their elders, to throw the useless burden away, so that in a very short time they all learnt to leave the beads alone.

Meanwhile another set of workers within the ant-city were busy stripping off the husks of the seeds

* *Atta structor* and *Atta barbara.*

and casting them out of the nest, and as in many cases whole seeds are either thrown out by mistake or dropped on their way in, one of these ants' nests may be found by noticing the little crop of oats, chickweed, and other grasses, which spring up round the refuse-heap.

By cutting a nest open or taking a good piece out of it with a trowel, the little granaries, in which these seeds are stored, may be laid bare. They are about the size of a gentleman's gold watch, and are connected by narrow galleries. It is a curious fact, that though these seeds grow easily when they are sown, yet in the granaries it is very rare to find one sprouting. This is probably owing to the care which the ants take to keep them dry, making the roof and sides of the granaries firm and hard, and bringing the seeds out on a warm day and spreading them round the nest, so that any moisture is drawn out of them. Mr. Moggridge even saw the ants, after a shower of rain had made the seed germinate, bite off the point of the little root which was beginning to show itself.

In this way the ants store up seeds in the summer, having often a large series of galleries and granaries, so that from half a pint to a pint of seeds has been taken from one nest; and in the winter, when food is scarce, the starch in these seeds supplies them with nourishment.

And now one vexed question still remains—have these clever little insects yet learnt to sow seeds as well as to gather them? This still remains to be proved; but if we travel to Texas we find that one thing is certain—namely, that they have learnt to

ANTS AND THEIR HELPLESS CHILDREN. 295

clear the ground round their nests, even among the toughest grass, and to allow nothing to spring up on these cleared disks except the needle grass or ant-rice,* which they store up in their granaries.

We have already seen that the English hill-ant will clear a path; but what labour it must be to clear, and keep clear, round spaces measuring from seven

Fig. 94.

Cleared disk of the agricultural ant, with a central mound and seven roads.—*Mc Cook.*

to twelve feet across, on wild meadow ground covered with rank weeds two or three feet high, some of them having stems as thick as one's finger! Yet this is done by the "agricultural or bearded ants"† of Texas, which swarm in such numbers and clear so many spaces that they actually injure the farms on which

* *Aristida stricta.*
† *Myrmica barbata (Pogonomyrmex barbatus).* H. C. McCook, Agricultural Ant of Texas. Philadelphia, 1879.

they establish themselves. They keep the circular space round their nests perfectly clean, never allowing a weed to encroach upon it except where at the edges crops of needle grass grow, of which they harvest the seed. Underground their galleries and chambers often extend under the whole disk, and there can be little doubt that it is chiefly in order to get air and ventilation, about which they are very particular, that they clear the weeds away. But their work does not end here, for they make from three to seven roads, according to the size of the nest, branching out into the forest of grasses, so that they can go far afield to collect seeds. These roads are often more than fifty feet long, and it sounds strangely like our own country places when we hear that they grow weedy in the winter when little used, and are cleared afresh in the spring.

When we think, however, of the small size of the ants in comparison with the vegetation they have to destroy, the history becomes much more astonishing. Many of the larger and thicker grass stems which they saw through with their mandibles to clear their disk must be to them like the trunks of trees measuring six feet across, while the round spaces they keep clear are, in relation to their size, equal to a piece of country a quarter of a mile in diameter.

These ants make their nests entirely underground, only sometimes having a small dome (see Fig. 94) with one or two openings in the top. Their granaries are very large, and yet they are not entirely vegetarians, for Mr. McCook saw them laying in a complete store of male and female termites which fell round their nest after swarming.

Central America can, however, boast of at least one purely vegetarian ant—namely, the "leaf-cutting ant."* These active little creatures devastate whole forests by tearing the leaves with their mandibles and carrying off pieces about the size of a sixpence into their nest, and Mr. Belt † found that these leaves are probably used for manure, upon which a minute fungus grows inside the nest and forms the ant food. These ants are decidedly clever, for when they were changing their nests once, and had to get their cocoons down a slope, Mr. Belt saw one set of workers bring them to the top and roll them down, while another set picked them up at the bottom. Another ant, which is housed and fed in a most peculiar manner, inhabits the Bull's-horn thorn-tree. This ant lives in the hollow thorns of the tree, sipping the honey which exudes from a gland at the base of the leaves, and in return, as it stings terribly, it protects its friend the tree from the attacks of the leaf-cutting ant.

The foraging or "army ants "‡ of Central America, on the other hand, subsist entirely on insects and other animal food, and travel in great hordes from place to place, clearing the country as they go, and living in hollow trees and fallen trunks on their road. Cockroaches, crickets, spiders, locusts, wood-lice, centipedes, and scorpions, all fall a prey to this huge moving army, often three or four yards wide, and the natives call it the "blessing of God," because the ants swarm into their houses, and by the time they leave every insect is cleared away. The army consists not only of dark workers and soldiers with

* Œcodoma.   † Naturalist in Nicaragua, 1874.   ‡ Eciton.

enormous heads and powerful jaws, but has also at intervals of about two or three yards light-coloured officers which touch the ants with their antennæ, and seem to give the orders directing the march. So the column moves on, each ant probably guided chiefly by scent and the other senses of its antennæ, for these ants are almost and sometimes entirely blind. Scouts are sent out on all sides to bring intelligence of booty, and the army swarms to the right or the left according to information given, following the scent of their comrades.

And now we must take leave of these intelligent little beings, though we have not even glanced at many of their curious habits, such, for example, as the storing up of honey in the abdomen of ants hanging from the roof of the nest, as is practised by the Mexican honey ant. But we have learnt enough to be convinced of their intelligence, and it only remains to inquire whether, amongst all their work, they have any feeling of sympathy for each other. The truth is, they seem to care for the members of their own nest, but more as parts of the community than as individuals. There are many cases in which ants have gone to help a comrade, but this is generally (though not always) when she is still able to share in their work; as, for example, when Mr. Belt tells us that the foraging ants never rested till they had released a comrade which he had covered over with a lump of clay. Sir J. Lubbock, it is true, gives one case of a poor ant born without antennæ, which was roughly handled by some enemies, and was afterwards most carefully carried home by a friend. But these incidents seem rare, and upon the whole the great guiding principle in ant-life appears to be devotion to the community, much more than to each other. With them the mother has no interest in her children after they are born, and the workers take care of all alike, so that no special ties of affection are formed; and, while we admire the wonderful mechanism of ant-life, we must not expect to find in it that love and personal devotion which is developed in quite another branch of Life's children.

We have travelled far since we started with the shapeless and sluggish Amœba, and have surely justified the statement with which we began, that by giving the prize of success to those who best fight the battle of existence, Life educates her children to fill their place in the world.

Much as we admire the tiny lime-builders and their beautiful shells, we must confess that the slime-animal itself is a frail and helpless being, with but feeble enjoyment of life, and the first advance which we perceive in the sponges is one rather of architecture than of individual existence. But in the lasso-throwers we already begin to detect the rudiments of those senses which afterwards become so keen; the nerves, eyes, and ears of the jelly-fish enable it at least to begin to appreciate the world around it and to live a free and independent life. In the star-fish and his companions we advance still further. Here is movement by walking as well as by swimming, a keen eye keeps a look-out on all things around, a battery of nerves, complicated muscles, and other parts give a far more distinct individuality and glimmerings of intelligence to the prickly-skinned animals than to the floating jelly, driven hither and thither by the currents of the sea. In fact these creatures stand at the head of a small but peculiar branch of life's children, while we have had to travel along another road to reach higher intelligence. This road led us from the oyster, so low in perception, yet so perfect in his internal mechanism, through a long chain of beings to the cunning octopus and cuttle-fish. Here we have the quick eye, the rapid movement, and the power of adapting things to the benefit of the animal, as when the little Sepiola blows a hole in the sand and arranges the stones round his body; we have the quick instinct of self-defence directing the inky fluid against an enemy, the capacity for changing colour for protection or attack, and the maternal care of the eggs. In a word, we arrive here at the head of one great division specially adapted for marine life, though some of its forms gain a footing upon the land. Still this division is incapable, so far as we can see, of advancing into successful competition with yet higher forms which, arising in the dim past almost from the same centre as these mollusca, have branched out on sea and land into crustaceans and insects. We need scarcely follow this branch through its ramifications, for the past chapters have shown the gradual progress of intelligence accompanying concentration of nervous power till we arrive at foresight, prudence, and organisation among the ants.

Still we must feel that something is wanting, and that something is mutual sympathy and help between any two beings, independently of mere duty as citizens. This we shall not find to any extent among the invertebrata or animals without back-bones, which are those we have dealt with in this book. Among the higher mollusca we find something like maternal care in the cuttle-fish; and the scorpion and earwig care for their young. But even among insects the large majority never live to see their children born, and those which do generally leave the care of them to others. We must turn for the development of fuller sympathy to that other branch, the key-note of whose existence is the relation of parents to children, of family love. If at a future time we are able to trace out the history of the vertebrate animals, it will be our great interest to watch the rise of this higher feeling. Then we may perhaps learn that the "struggle for existence," which has taught the ant the lesson of self-sacrifice to the community, is also able to teach that higher devotion of mother to child, and friend to friend, which ends in a tender love for every living being, since it recognises that mutual help and sympathy are among the most powerful weapons, as they are also certainly the most noble incentives, which can be employed in fighting the battle of life.

INDEX.

ABDOMEN, definition of term, 156.
Acineta or tube-sucker, 21.
Acontia or darts of the anemone. 70.
Acorn barnacle, structure of the, 174, 175.
Actinozoa, 55.
Africa, scorpions of, 180.
Africa, termite mounds of, 230.
Air-thimble of water-spider, 197.
Air-tubes of insects, 212.
America, harvesting ants of, 293-295; scorpions of South, 180.
Amœba, feeding, 18.
Anemone, section of an, 67.
Anemones of the sea, 66; group of, 68; their rank among animals, 10; birth of young, 69; food and enemies of, 69; lasso-cell of, 53; parasitic on hermit-crab, 172.
Animal, first walking, 89; the simplest, 16.
Animals, distribution of, 7; which change their form during life, 234.
Animal-trees, 56.
Ants, agricultural, 295; foraging, 297; garden, 277-287; harvesting, 293; hill or horse, 271-277; leaf-cutting, 297; meadow, 287; negro, 288-291; red, 288; living in hollow thorns, 297; slave-making, 288; antennæ-language of, 274; and aphides, 270-278; capturing termites, 296; cocoons of, 280; eggs of, 279; food of, 276; friendliness of, 298; manner of digging, 275-283; nests of, 279, 283, 295; number in one nest, 270; migrations of, 283; points of resemblance to man, 270; recognition by, 287; road-making of, 284, 296; rolling cocoons down a slope, 297; slave-making expeditions, 289; stinging and stingless, 272; structure of, 272; wars of, 285; winged, 282; nervous system of the, 276; young needing more help than the bee, 267; feeding each other, 278.
Antedon (Comatula) rosacea, 90.
Antennæ of insects, 156.
Antennules of prawn, 160.
Ant-lion, funnel of the, 225.
Aphides, 201; eggs protected by ants, 287, 292; multiplication of, 203; winged, 203; their relation to ants, 278, 292.
Aphrodite or sea-worm, 151; hispida, harpooned bristles of, 152.
Apple-trees destroyed by aphides, 205.
Arctic regions, small crustaceans of, 160.
Argonaut, figure of, 132; does not

sail, 131; male has no shell, 132.
Argyroneta aquatica, 196.
Aristotle on mouth of sea-urchin, 97.
Arthropoda or jointed-footed animals, 155.
Ascidians, 103 *note*, 117 *note*.
Atlantic telegraph, mud from the, 28.
Atta structor and Atta barbara, 293.

BASKET-FISH, 92.
"Balancers" of the daddy-long-legs, 261.
Balanus, structure of the, 176.
Barnacles, floating (Lepas), 174.
Bate, Mr. Spence, on acorn barnacle, 176; on hearing of prawn, 161.
Beads taken by mistake by ants, 293.
Beetles, adaptation of parts to work, 261; undergo metamorphosis, 251; muscular power of, 260; parasitic, 260; rove, 251; cockchafer, 252; plant-eating, 252, 255; carnivorous, 256; water, 257; cocktail, 256; bombardier, 255; whirligig, 257; blind, 260; bee, 260; skipjack, 261; sexton, 259; goliath, 251; tiger, 255; carnivorous, 257.
Bees, organisation of, 268.
Bell-flower or Vorticella, 21, 31.
Belt, Mr., on leaf-cutting and foraging ants, 297; on ants helping each other, 298.
Bivalve shells, formation of, 106.
Birds, rate of increase of, 4.
Black-ant nest, attack on, 289.
Bladder-worm, 139.
Bluebottle's early life, 263.
Botflies, their manner of wounding, 263.
Bowerbank, Dr., on sponge spicules, 45 49.

Brachiopoda, 103 *note*.
Brain-coral, 75.
Breathing of spider, 188; of dragon-fly grub, 223; of gnat grub, 264; of sea-mouse, 151.
Breathing-chamber of land snails, 120; -holes of a caterpillar, 238.
British Museum, shells in the, 106; forms of star-fish in, 101.
Brittle star-fish, infancy of a, 79; full-grown, 84-91; its movements and habits, 92.
Brobdingnag and Gulliver, 269.
Bugs, air and water, 207.
Bunodes gemmacea, 68.
Burnet-moth with caterpillar and cocoon, 246.
Butler, Mr. A., on spider's web, 187.
Butterfly, life from a caterpillar, 236-240; head and egg of a, 237; formation of perfect, 240.
Butterflies and moths, comparison of, 243.

CABBAGE-BUG, Pentatoma, 206.
Cabbage butterfly's mode of binding the chrysalis, 243.
Caddis-fly and grub, 220-222.
Calamary, a huge arm of, 131.
Calamaries, horny pen and hooked suckers of, 130.
Campanulina, 59.
Cardium, 111.
Carinaria atlantica, 125.
Carter, Mr., cited, 39.
Caryophyllium Smithii, 75.
Caterpillar, head and foot of, 237; life of a, 236-239; leaf-rolling, 247; goat-moth, 248; Psyche in case, 247; Burnet-moth, 246.
Cases of the caddis-grub, 222.
Cave anemone, 68.
Centipede, 156.
Cephalopoda, or head-footed animals, 128.

INDEX. 305

Ceratium, Protogenes feeding on a, 17.
Cerceris, paralysing insects for food, 267.
Ceylon, land leeches of, 143.
Chalk formed of foraminifera, 28.
Chalk-beds, extent of, 28.
Chitine, nature of, 57, 157.
Chrysalis bound to a paling, 243; emerging from caterpillar skin, 240, 243.
Chysoara hysocella, figure of, 63.
Cilia, or whip-like lashes, 38, 64; on gills of the oyster, 109.
Cirrhi, on legs of acorn-barnacle, 175.
Claws, snapping, of the star-fish, 88; of sea-urchin, 98.
Cleanliness of spiders, 197; of ants, 284.
Clothes-moth, history of the, 249.
Coal-mines, insect remains in, 210.
Cochineal insect, 207.
Cockchafer, grub, cocoon and beetle, 252.
Cockle, figure of, 111; siphons of, 113; leaping foot of, 113.
Cockroaches, clever escape of, 217; figure of, 216; enemies of, 218.
Cocoon carried by spider, 195; how the moth emerges from a, 245.
Cocoons of ants, 280; of spiders, 190, 194; of moths, 245; leaf, 247; hairy, 247; rolled down a slope by ants, 297.
Cœlenterata, 55.
Coleoptera, term explained, 251.
Colour, changing of octopus, 129; in mantle of mollusca, 105.
Comatula (Antedon) rosacea, 90.
Coral, history of red, 71; section of, 73; growth of white, 74.
Crab, development of a, 167-169; changing his shell, 169; mate watching a soft, 170; common, 167; fiddler-crab, 172; racing-,

173; frog-, 173; robber-, 173; land-, 173; carrying a sponge, 38.
Cray-fish, 159.
Crickets, night insects, 215.
Crinoids or stone-lilies, 78.
Crops of grass round ants' nests, 294-296.
Crustacea, 154; of arctic regions, 160; their rank among animals, 11.
Crustacean parasites, 177.
Crustaceans, various forms of, 176.
Cuckoo spit insect, 205.
Cup-coral of Devonshire, 75.
Cuttle-fish, eggs and "bone" of, 130.
Cypris, freshwater crustacean, 177.

DADDY-LONG-LEGS, showing his balancers, 262; ovipositor of mother, 263.
Daisy anemone, 68; young of, 69.
Dalyell, Sir J., on sea-cucumber, 100.
Darwin, Mr., on octopus taking aim, 128; on structure of acorn-barnacle, 176.
Devonshire cup-coral, 75.
Diptera, or two-winged flies, 261.
Disk of agricultural ant-nest, 295.
Distoma militare, 138.
Divisions of animal life, 10.
Doris pilosa, a sea-slug, 124.
Dragon-fly, life of the, 222-225.
Dyticus marginalis, true water-beetle, 256.

EARTHWORM, structure and habits of, 146; cocoons of the, 147.
Earwig mother, 218.
Echinodermata, 82.
Echinus, tube-feet of, 94.
Ecitons or hunting ants, 297.
Education of life, 6.
Edwardsia calimorphia, 68.
Eggs of ants, 279; of aphides cared for by ants, 292; of

campanulina carried in a jelly-bell, 60; of cockroach in a case, 217; of cuttle-fish, 130; of argonaut, 131; of octopus, 131; of snails and slugs, 122; of spiders, 190-194.
Elastic-ringed animals, 135.
Elytra of beetles, 251.
Encrinites or fossil stone-lilies, 79.
Enemies of the sponge, 43.
Eolis coronata, a sea-slug, 124.
Emerton, Mr., on spider's web, 187.
Ephemera, life of the, 221.
Erber watching trapdoor spiders, 193.
Eriosoma lanigerum, apple aphis, 205.
Eyes of ants, 273; of caterpillar and butterfly, 241; of prawn, 161; of the star-fish, 88; of the sea-urchin, 96; of the snail, 120; of whirligig beetles divided, 258; of the young oyster, 110; of a young crab, 168.

FABRE, M., on dung-feeding beetles, 259.
Fairy-shrimps, gills of, 177.
Feather star-fish, infancy of a, 78-90; full-grown, 89, 90.
Feet, tube-, of echinodermata, 80-94; true and false of a caterpillar, 239.
Fiddle-crab seizing the hermit, 171.
Fire-flies, 260.
Five-fingered star-fish, 80-84.
Flea, early life of a, 234.
Flesh-feeding molluscs, 118.
Flies, various two-winged, 262.
Flint-shells, 29, 30.
Flint-sponges, 46.
Fly, house-, where hatched, 263; size of brain of, 265.
Flukes or flat worms, 138.
Foraging ants, 297.
Foraminifera, definition of name, 23; growth of perforated, 27.

Forbes, Prof., on star-fish's wink, 89; on contortions of brittle star, 89.
Forel, M., on ant migrations, 284.
Formic acid used in ant-battles, 285.
Formica fusca, negro-ant, 288; -rufa, 277; -sanguinea, slave-making ant, 288.
Frog-hopper insect, Aphrophora, 206.

GADFLY, early life of, 263; food of the two sexes, 262.
Ganglia of leech, 144.
Garden ant lives underground, 277; keeps aphides, 287; its mode of fighting, 288.
Gaucho or lasso-thrower, 51.
Gem-pimplet, 68.
Gerris, a water-measurer, 208.
Gills of fairy shrimps, 177; of land-crabs, 173; of May-grub, 221; of octopus, 127; of nautilus, 133; of oyster, 108; of sea-slugs, 124; of skeleton shrimp, 163; of prawn, 164.
Gizzard of grasshopper, 213.
Glass-rope sponge, 46.
Globigerina, 23, 27; in chalk, 28.
Glow-worms, 260.
Gnat, history of the, 263-265.
Goat-moth caterpillar, 248.
Goliath-beetle, 251.
Gossamer webs, 196.
Gosse, Mr., on movement of scallop, 112.
Granaries in ant-hills, 294.
Grass crops on ants' nest, 294, 295.
Grasshopper, early origin of the, 210; large green, and young, 211; spiracles of, 212; gizzard of, 213; little green, is a locust, 214; general structure of, 211; laying her eggs, 211, 214; cry of the, 215.
Grecian Archipelago, sponges of, 35.

INDEX. 307

Grub feeding on aphides, 202-204; of dragon-fly feeding, 222.
Gulf of Mexico, sponges of, 35.
Gulliver cited, 269.

HAECKEL on eye of star-fish, 88; finding Protogenes, 16; cited, 18, 29, 156.
"Hanging-bell" jelly-fish, 66.
Harvesting ants, 293.
Harvest-bug a mite, 199.
Head of ant, 273.
Headless mollusca, 111.
Heads, absence of, in lower animals, 102.
Helplessness of infant hymenoptera, 266.
Hemiptera, term explained, 209.
Hermit-crab, parasites of, 172; structure and habits of, 170.
Hertwig on nerves of medusæ, 61.
Hill-ant, structure of, 271; nests of, 277.
Holdsworth, Mr., on birth of anemones, 69.
Honey-stealing caterpillars, 250.
Honey-tubes of aphides, 203.
Hop harvest destroyed by aphides, 205.
Horse, gad-fly hatched inside the, 263; -mussel, pea-crab in the, 136.
Huber on slave-making ants, 291, 292.
Hunting ants, 297.
Huxley on rate of increase of plants, 4; on structure of acorn barnacle, 174.
Hydra, figure of, 51; food of, 52; lasso-cells of the, 53.
Hydra form of jelly-fish, 64.
Hydrophilus or black water-beetle, 256.
Hydrozoa, 55.
Hymenoptera, 263.

IANTHINA or ocean-snail, 125.

Ichneumon fly placing eggs, 267.
India, termites or white ants of, 225.
Indian Ocean the home of the nautilus, 133.
Infusoria and their origin, 20.
Ink-bag of octopus, 127.
Insects, air-tubes of, 213; at the head of invertebrates, 11; complete metamorphosis of, 233; paralysed for food, 267; proportion of, among animals, 158; use of term, 155.
Insect's eye, section of an, 224.
"Insects of the sea" or crustacea, 156-159.
Intelligence of ants, 298; of hymenoptera, 266.
Invertebrata, or animals without backbones, 300.
Ireland, feather-stars of, 91.
Italian markets, sea-urchins in, 98.
Italy, coral on coasts of, 73.

JELLY-BELLS, 55, 59, 61.
Jelly-fish, their rank among animals, 10, 54; small weight of solid matter in, 62; food of, 63; birth and childhood of the, 65.
Johnstone, Dr., on sponges, 39.
Jointed-footed animals, 155.

King-crabs, 177.
Kingdoms, animal and vegetable, 10.
Knobs, in stalk of ant's body, 271.

LAC insect, 207.
Lady-bird, 256.
Lagena 23.
Land-snails, breathing of, 120.
Lankester, Mr. Ray, on insects, 156.
La Rochelle damaged by termites, 230.
Lasius niger, 272.
Lasso-cells of the hydra, 52.
Lasso-throwers, meaning of term, 51; various forms of, 54.

Leaf and stick insects, 218.
Leaf-miners and their cocoons, 248.
Leaf-rolling caterpillars, 248.
Leg of ant bearing combs, 273.
Leech, food and young of, 145; nervous system of, 143-145.
Leeches, land, of Ceylon, 143.
Lepidoptera, explanation of term, 240.
Life, various forms of, 2; rapid increase of, 4.
Lima building a nest, 112.
Limax maximus, 28,000 teeth of, 121.
Lime-sponges, figure of, 44.
Limpet, figure of, 114; habits of, 117.
Limulus or king-crab, 177.
Lingthorn's eye winking, 89.
Linnæus, on division "insecta," 155.
Liver-fluke, 139.
Lobster, breathing-gills of the, 164.
Lobsters, number sold in London, 159; rapid multiplication of, 172.
Locust-swarms, 214.
Long-worm, Nemertes Borlasia, 137.
Lowne, Mr., on size of fly's brain, 265; on spider's web, 187.
Lubbock, Sir J., on ant-communication, 274; on sorting of ant-grubs, 280; on age of ants, 281; on ant-queen working alone, 282; on ants kept alive by slave, 292; on ants storing aphis eggs, 292; on kindness of ants; 298; on recognition by ants, 287.
Lugworm, 150.
Lyonnet on air-tubes of insects, 213.

McCook on Texas ants, 295, 296.
Madrepore coral, 74.
Madreporiform tubercle, 87.
Maggot of nut, 254; of pea, 255.
Mandibles of ants, uses of, 275.

Mantis or snatching insect, 218.
Mantle of mollusca secreting shell, 104.
"Mask" of dragon-fly grub, 222.
May-bug or cockchafer, 252.
May-flies do not eat, 220; and their grubs, 219-221.
Meadow-ant keeping aphis eggs, 287.
Mediterranean, coral of the, 55, 71; harvesting ants of the, 293; large octopuses of the, 131; sea-urchins of, used for food, 98; scorpions of the, 180; trap-door spiders of the, 193.
Medusæ, freshwater, 54.
Medusa's head, 78.
Membrane-winged insects, 266.
Metamorphosis imperfect in cockroach, 235; of crab, 169; of gnat, 264; of insects, 233.
Mermis, a thread-worm, 140.
Mexico, cochineal insect of, 207.
Migrations of ants, 283.
Miliolite forming its shell, 24.
Miliolites, birth of young, 25.
Mites, land and water, 198; parasitic, 199.
Moggridge, Mr. T., on harvesting ants, 293; on seed bitten when sprouting, 294.
Mollusca, meaning of term, 104; shell-secreting mantle of, 104; naked-gilled, 123; possible relationship to worms, 134; their rank among animals, 11; vegetable-feeding, 114; flesh-feeding, 118.
Monads, their origin, 20, 31.
Money-spinners, 196.
Mosquito, mouth of, 263.
Mother-of-pearl, cause of, 106.
Moth, sphinx-, 244; silkworm-, 245; oak-eggar-, 246; Burnet-, 246; procession-, 247; Psyche-, 247; goat-, 248; clothes-, 249.
Moths and butterflies, comparison of, 243; and their cocoons, 245-248.

INDEX.

Myrmica molesta, 271.
Mytilus, figure of, 111.
Myzoxyle mali, apple-aphis, 205.
Murie, Dr., cited, 41.
Mushroom, tiny beetles in, 255.
Mussels, anchoring-threads of, 111.

NAKED-GILLED mollusca, 123.
Nautilus, structure of, 133.
Nereis, a sea-worm, 151.
Nerve-winged insects, 219.
Nervous system of medusæ, 61; of star-fish, 86; of mollusca, 109; of leech, 144; of spider, 189; of prawn, 165; of a caterpillar, 238; of ants, 276.
Nests of ants, 279; formation of new, 283.
Noctiluca, or night-glow, 15, 19, 21.
Neuroptera, early origin of, 219.
Newport, Mr., on metamorphosis, 236.
Nodosarina, 23.
Nummulites forming the Alps, 27.
"Nurses" of flukes, 138.
Nurses helping young ant, 281.
Nut-weevil, 254.

OAK-EGGAR moth, 246.
Ocean-snail, Ianthina, 125.
Octopus shooting backwards, 127; complicated structure of, 126; inky fluid of, 128; suckers in arms of, 129; mother and eggs, 131; changing colour of, 129.
Œconoma, or leaf-cutting ant, 297.
Operculum of mollusca, 105.
Ophiocoma bellis, 84.
Orang-outang, helplessness of young, 266.
Orbitolite shells, 23, 26, 28.
Orthoptera, term explained, 210.
Ostrea edulis, figure of, 108.
Outcasts of animal life, 135.
Ovipositor of grasshopper, 211; of bot-fly, 263.

Oyster, infancy and perils of the, 109; structure of the, 107, 108.
Oyster-beds, 107.

PACIFIC, coral islands, 55, 71, 74.
Painted Pufflet, 68.
Parasites, 136, 141, 177, 199, 260; degradation of, 141.
Parasitic beetles, 260; mites and ticks, 199.
Paris houses built of orbitolite limestone, 28.
Pea-crab, a parasite of the horse-mussel, 136.
Pea-maggot, 255.
Pearls, how formed, 106.
Pecten, 111.
"Pen" of calamaries, 130.
Penerepolis, 23.
Pentacrinus caput-medusæ, 78.
Periwinkle, formation of shell of, 105; inside of a, 115; toothed rasp of the, 116; gills of the, 117.
Pholas, burrowing habits of the, 113.
Phosphorescence, caused by jelly-bells, 61; from one jelly-fish, 64; of flies and glowworms, 260; on the sea, 15, 21.
Phosphoridæ, 66.
Phryganea or caddis-fly, 221.
Physematium, 29.
Pincers of scorpion, 179.
Pinna, anchoring threads of, 112.
Planaria, 137.
Plants, rate of increase of, 4.
Plant-bugs, 207.
Plant-lice, 202.
Plates of brittle-star, 92; of sea-urchin, 95-97.
Poison-dart of scorpion, 179; -fangs of spider, 183-188.
Polycistinæ or flint builders, 30.
Polyergus rufescens, mode of fighting, 288; helplessness of, 291.
Polypites, nature of, 57.

LIFE AND HER CHILDREN.

Polyzoa, not dealt with, 103, *note.*
Portuguese man-of-war, 66.
Prawn, structure of the, 160-164; shedding his skin, 165; cleaning himself, 166.
Prickly-skinned animals, 82.
Princesses among ants, 282.
Procession-moths and their cocoons, 247.
Protamœba, 18.
Protogenes, or thread-slime, 16.
Protozoa, 31.
Psyche-caterpillars in tubes, 247.
Pteropods or wing-footed snails, 126.
Pupa of a butterfly, 240.
Pyramids formed of nummulite limestone, 28.

QUEEN ANTS laying eggs, 279; no jealousy between, 282.
Queen-termite, 228.

RADIATE plan of structure, 96.
Radiolariæ or flint-builders, 30.
Rasp (radula) of the periwinkle, 116.
Rays, of star-fish, 85; of brittle-star, 90; of sea-urchin, 95; of sea-cucumber, 99.
Razor-fish, figure of, 111; burrowing habits of, 113.
Red coral, growth of, 72.
Red mite of vines, 199.
Red Sea, sponges of, 35.
Ringed bodies of insects, 155.
Roads made by ants, 284.
Romanes, Mr., on nerves of medusæ, 61.
Rose, group of aphides on a, 202.
Rosy feather-star, 89.
"Rot" in sheep, 139.
Rotalia, 23-27.
Rotifera, 137.

SAGARTIA viduata, S. bellis, and S. troglodytes, 68.
Sagartiaœæ, special darts of the, 69.
Sand-hopper, Talitrus, 163.

Sand-wasp, paralysing insects for food, 267.
Saw-flies, young, 267.
Scallop, figure of, 111; eyes of the, 112.
Scarabæus beetle, 258.
Scent of moths attracting mates, 250.
Schäfer, on nerves of medusæ, 61.
Schultze, on birth of miliolites, 25.
Scorpion, structure of, 180; figure of with cricket, 179.
Sea-anemone. *See* anemone.
Sea-cucumber, infancy of a, 82; power of regrowth in the, 10, 100; structure and food of, 99.
Sea-fir, Sertularia cupressina, 58.
Sea-mouse or Aphrodite, 150.
Sea-nymphs, 126.
Sea-oak coralline, figure of, 56.
Sea-slugs, 123; figures of, 124; food of, 124.
Sea-urchin, infancy of a, 81; walking on a rock, 94; stripped of its spines, 95; wrapped in sea-weed, 93; structure of a, 95; growth of shell of, 96; mouth of, 97; food of, 97; tube-feet of, 94.
Sea-worms, fixed, 148; active, 151.
Seeds collected by harvesting ants, 293.
Sepiola blowing a hole in sand, 130.
Serpula, its tube and tentacles, 148.
Sertularia cupressina, 58; S. plumula, 56.
Sexton or "burying" beetles, 259.
Sheath-winged insects, 251.
Shell of argonaut, a cradle, 132; of sea-urchin, 96.
Shell-builders, the simplest, 22.
Shrimp, hand of the, 163.
Silkworm, how it spins, 245.
Simplest children of life, 10, 14.
Siphon of octopus, 127.
Skeletons of sponges, 43-49.

INDEX. 311

Skeleton, outside, of insects, 54, 157.
Skeleton shrimp, Caprella, 163.
Slave-making ants, 288; helplessness of one kind, 291.
Slug, figure of, 122; hidden shell in back of, 121.
Smeathman on termites, 228, 231.
Snails, metamorphosis of worms in, 138; winter shelter of, 121; eyes and breathing chamber of, 120.
Snake-locked anemone, 68.
Snare-weavers, 179-200.
Solen, 111.
South America, calamary's arm from, 131.
Sphex paralysing insects for food, 267.
Sphinx moths and their caterpillars, 244.
Spicules of sponges, 45.
Spider, nervous system of, 189; males feeble and small, 190; cocoons of, 190, 191, 195; structure of, 183; entangling her victim, 188; manner of spinning web, 184.
Spiders, house, web of, 185; hunting-, 195; tunnelling-, 191; water-, 196; trap-door-, 192; gigantic, 197.
Spines of sea-urchin, 94, 97.
Spinnerets of spider, 184.
Spiracles of grasshopper, 212.
Spirorbis, 148.
Sponge, British, 37; flint, 47; section of, magnified, 46; lime, 44; cup, 48; rank of, 35, 41; spicules of, 45; homes of, 35; section of bath-, 41; -fisheries, 36.
Sponge-animal, growth of, 39-42; eggs of, 37, 38; flesh composition of, 35; tissue, 34.
Sponges, boring, destroying oysters, 110.
Spur on leg of ant, 271-273.

Squids, ten-armed cephalopods, 130.
Star-fish, eyes of the, 88; infancy of common, 80; its rank among animals, 10; various forms and sizes of, 101; walking apparatus of the, 83-86; figure of common, 84; section of a, showing structure, 85; food of the, 87; water-hole of the, 85.
Staveley, Miss, on burying beetles, 259.
Stone-lily the young of feather-star, 91.
Stone-lilies or crinoids, 78.
Straight-winged insects, 210.
Struggle for existence, 5-7, 12.
Suckers in arms of octopus, 129.
Sun-slime, 29.
Sympathy, how far existing, in ants, 298.
Synapta a kind of sea-cucumber, 101.

TEETH in lobster's stomach, 168; of mollusca, 116.
Tennent, Sir E., on Ceylon leeches, 143.
Tentacles of hydra, 53; of sea-cucumber, 83; of snail, 120.
Terebella, or shell-binding worm, 148, 149.
Teredo, a mollusc, 113.
Termites, or white ants, 225-231; figures of, 226; mode of working of, 227, 229; queen-cell of the, 228; eggs and nurseries of the, 229; marching columns of, 231; captured by ants, 230.
Termite mounds of Africa, 230.
Testacella, figure of, with shell, 123.
Texas, agricultural ant of, 295.
Textularia, 23.
Thorax, definition of term, 156; of grasshopper, 212.
Thread-slime or Protogenes, 16.
Ticks, 136.

Tiger-beetle grub feeding, 256.
Tools of an animal grow upon it, 7.
Tortoise-shell butterfly, life of the, 236.
Trachea or breathing-tube, 212.
Trap-door spiders, 192.
Trepangs, 101.
Trichina in pork, 140.
Trilobites, 173.
Tube-feet of star-fish, 80; of sea-urchin, 94.
Tube-hydra or Tubularia, 58.
Tube-sucker or Acineta, 21.
Turkey, sponges of, 49.
Two-winged flies, 261.

UNIVALVE SHELLS, formation of, 106.
Urastei rubens, 84.

VEGETABLE-FEEDING MOLLUSCA, 114.
Venus' Basket, figure of, 47.
Vertebrata and their divisions, 12.
Vesicles, supplying the tube-feet of star-fish, 87.
Victoria Regia, Medusa in tank of, 54.
Vorticella or bell-flower, 21.

WALLACE on rate of increase of birds, 4; cited 266.
Wars of ants, 285.
Wasp, spider releasing a, 189.
Wasps, nests of, 267.
Water-boatman, Notonecta, 208.
Water-cresses, flat-worms on, 137.

Water-flea (Daphne), 159, 177; -measurers, 207; -mites, 136, 199; -snails, 120; -spider, 196.
Weapons of animals, meaning of term, 7; of sea-worms, 150.
Web of garden spider, 185; of house spider, 191; of tunnelling spider, 191.
Weevils, 254.
West Indies, land-crab of the, 173.
Wings of bugs different from beetles, 209; of butterflies, 240.
Winged ant killed by slave-makers, 291.
Winged ants, 282.
Wing-footed snails, 126.
Whale feeding on jelly-fish, 76.
Whelk and eggs, figure of, 118; drilling rasp of, 119; young of, free-swimming, 119.
Whelk-shell, hermit crab in a, 170.
Whip-cells of a sponge, 40.
White ants, not true ants, 225, 231.
White, Mr. Charters, on a fish within a sea-anemone, 69.
Wood-louse a crustacean, 173.
Worms, their rank among animals, 11, 135; parasitic, 138-141; sea-, 151; ribbon-, wheel-, and long-, 137.
Workers of ants imperfect females, 276.

YELLOW ant tending eggs of aphides, 287.

ZYGENA filipendula, 246.

THE END.

BOOKS FOR YOUNG READERS.

THE WINNERS IN LIFE'S RACE; or, The Great Backboned Family. By ARABELLA B. BUCKLEY. With numerous Illustrations. 12mo. Cloth, gilt, $1.50.

CONTENTS.—I. The Threshold of Backboned Life; II. How the Quaint Old Fishes of Ancient Times have Lived on into our Day; III. The Bony Fish, and how they have spread over Sea, and Lake, and River; IV. How the Backboned Animals pass from Water Breathing to Air Breathing, and find their Way out upon the Land; V. The Cold-Blooded Air-Breathers of the Globe in Times both Past and Present; VI. The Feathered Conquerors of the Air.—Part I. Their Wanderings over Sea and Marsh, Desert and Plain; VII. The Feathered Conquerors of the Air.—Part II. From Running to Flying, from Mound-Laying to Nest-Building, from Cry to Song; VIII. The Mammalia or Milk-Givers, the Simplest Sucking Mother, the Active Pouch-Bearers, and the Imperfect-Toothed Animals; IX. From the Lower and Small Milk-Givers which find Safety in Concealment, to the Intelligent Apes and Monkeys; X. The Large Milk-Givers who have Conquered the World by Strength and Intelligence; XI. How the Backboned Animals have Returned to the Water, and Large Milk-Givers Imitate the Fish; XII. A Bird's-Eye View of the Rise and Progress of Backboned Life.

"Although the present volume, as giving an account of the *vertebrate* animals, is a natural sequel to, and a completion of, my former book, 'Life and her Children,' which treated of *invertebrates*, yet it is a more independent work, both in plan and execution, than I had at first contemplated. . . . I have endeavored to describe graphically the early history of the backboned animals, so far as it is yet known to us, keeping strictly to such broad facts as ought in these days to be familiar to every child and ordinarily well-educated person: if they are to have any true conception of Natural History. At the same time I have dwelt as fully as space would allow upon the lives of such modern animals as best illustrate the present divisions of the vertebrates upon the earth: my object being rather to follow the tide of life, and sketch in broad outline how structure and habit have gone hand in hand in filling every available space with living beings, than to multiply descriptions of the various species."—*From the Preface.*

"An account of vertebrate animals, written with such natural spirit and vivacity, that it might convert even a literary person to natural science."—*Saturday Review.*

"We can conceive no better gift-book than this volume. Miss Buckley has spared no pains to incorporate in her book the latest results of scientific research. The illustrations in the book deserve the highest praise; they are numerous, accurate, and striking."—*London Spectator.*

"It is full of instructive illustrations."—*New York World.*

LIFE AND HER CHILDREN. Glimpses of Animal Life from the Amœba to the Insects. By ARABELLA B. BUCKLEY. With upward of One Hundred Illustrations. 12mo. Cloth, $1.50.

CONTENTS.—I. Life and her Children; II. Life's Simplest Children; how they Live, and Move, and Build; III. How Sponges Live; IV. The Lasso-Throwers of the Ponds and Oceans; V. How Starfish Walk and Sea-Urchins Grow; VI. The Mantle-Covered Animals, and how they Live with Heads and without them; VII. The Outcasts of Animal Life, and the Elastic-ringed Animals by Sea and by Land; VIII. The Mailed Warriors of the Sea, with Ringed Bodies and Jointed Feet; IX. The Snare-Weavers and their Hunting Relations; X. Insect Suckers and Biters, which Change their Coats, but not their Bodies; XI. Insect Gnawers and Sippers, which Remodel their Bodies within their Coats; XII. Intelligent Insects with Helpless Children, as illustrated by the Ants.

"The main object is to acquaint young people with the structure and habits of the lower forms of life; and to do this in a more systematic way than is usual in ordinary works on natural history, and more simply than in text-books on zoölogy. For this reason I have adopted the title, 'Life and her Children,' to express the family bond uniting all living things, as we use the term 'Nature and her Works' to embrace all organic and inorganic phenomena: and I have been more careful to sketch in bold outline the leading features of each division than to dwell on the minor differences by which it is separated into groups.'—*From the Preface.*

[SEE NEXT PAGE.]

BOOKS FOR YOUNG READERS.

THE FAIRY-LAND OF SCIENCE. By ARABELLA B. BUCKLEY. With numerous Illustrations. 12mo. Cloth, $1.50.

"A child's reading-book, most charmingly illustrated, and in every way rendered especially interesting to the juvenile reader."—*Athenæum.*

"So interesting that, having once opened it, we do not know how to leave off reading."—*Saturday Review.*

"Her methods of presenting certain facts and phenomena difficult to grasp are most original and striking, and admirably calculated to enable the reader to realize the truth. As to the interest of her story, we have tested it in a youthful subject, and she mentioned it in the same breath with 'Grimm's Fairy Tales.' . . . The book abounds with beautifully engraved and thoroughly appropriate illustrations, and altogether is one of the most successful attempts we know of to combine the *dulce* with the *utile*. We are sure any of the older children would welcome it as a present; but it deserves to take a permanent place in the literature of youth."—*London Times.*

"A child's reading-book admirably adapted to the purpose intended. The young reader is referred to nature itself rather than to books, and is taught to observe and investigate, and not to rest satisfied with a collection of dull definitions learned by rote and worthless to the possessor. The present work will be found a valuable and interesting addition to the somewhat overcrowded child's library."—*Boston Gazette.*

"Written in a style so simple and lucid as to be within the comprehension of an intelligent child, and yet it will be found entertaining to maturer minds."—*Baltimore Gazette.*

"The ease of her style, the charm of her illustrations, and the clearness with which she explains what is abstruse, are no doubt the result of much labor; but there is nothing labored in her pages, and the reader must be dull indeed who takes up this volume without finding much to attract attention and to stimulate inquiry."—*Pall Mall Gazette.*

SHORT HISTORY OF NATURAL SCIENCE AND THE PROGRESS OF DISCOVERY, from the Time of the Greeks to the Present Day. For Schools and Young Persons. By ARABELLA B. BUCKLEY. With Illustrations. 12mo. Cloth, $2.00.

"The volume is attractive as a book of anecdotes of men of science and their discoveries. Its remarkable features are the sound judgment with which the true landmarks of scientific history are selected, the conciseness of the information conveyed, and the interest with which the whole subject is nevertheless invested. Its style is strictly adapted to its avowed purpose of furnishing a text-book for the use of schools and young persons."—*London Daily News.*

"A most admirable little volume. It is a classified *résumé* of the chief discoveries in physical science. To the young student it is a book to open up new worlds with every chapter."—*London Graphic.*

"The book will be a valuable aid in the study of the elements of natural science."—*Journal of Education.*

"Miss Buckley supplies in the present volume a gap in our educational literature. Guides to literature abound; guides to science, similar in purpose and character to Miss Buckley's History, are unknown. The writer's plan, therefore, is original, and her execution of the plan is altogether admirable. She has had a long training in science, and there are signs on every page of this volume of the careful and conscientious manner in which she has performed her task."—*Pall Mall Gazette.*

For sale by all booksellers; or sent by mail, post-paid, on receipt of price.

New York: D. APPLETON & CO., 1, 3, & 5 Bond Street.

BOOKS FOR YOUNG READERS.

FACTS AND PHASES OF ANIMAL LIFE, interspersed with Amusing and Original Anecdotes. By VERNON S. MORWOOD, Lecturer to the Royal Society for the Prevention of Cruelty to Animals. With Seventy-five Illustrations. 12mo. Cloth, gilt side and back, $1.50.

CONTENTS: CHAP. I. Wonderful Facts about Animals; II. At the Bottom of the Sea; III. A Hunt in our Ditches and Horse-Ponds; IV. Buzzings from a Beehive; V. Spinners and Weavers; VI. Black Lodgers and Miniature Scavengers; VII. Insects in Livery, and Tiny Boat-Builders; VIII. Our Birds of Freedom; IX. Our Feathered Laborers: their Work and Wages; X. In the Building Line; or, Bird Homes and Family Ties; XI. Bird Singers in Nature's Temple; XII. Chanticleer and his Family; XIII. Miners of the Soil; XIV. Active Workers, with Long Tails and Prickly Coats; XV. Nocturnal Ramblers on the Lookout; XVI. Quaint Neighbors and their Shaggy Relations; XVII. Our Furry Friends and their Ancestors; XVIII. Our Canine Companions and Tenants of the Kennel; XIX. Relationship of Man and Animals; XX. Can Animals Talk and Reason? XXI. Useful Links in Nature's Chain; XXII. Clients worth Pleading for; Glossary and Index.

"'Facts and Phases of Animal Life' is a rare book in the natural history of the more common animals, and of a character very desirable for circulation to promote knowledge and love of animals. It gives wonderful facts about animals; tells of the structure and habits of those at the bottom of the sea, in ditches and horse-ponds; of bees, spinners and weavers, black lodgers and miniature scavengers; insects in livery and tiny boat builders; birds of freedom, feathered laborers, bird homes and family ties, bird singers, fowls; miners of the soil, active workers with long tails and prickly coats, etc., etc. It is well written and is very complete in its facts, quite sufficiently so for an introduction to the study of natural history, or for the general reader. It is of the better class of books for youth, and will benefit them by telling them a good deal that they ought to know, and teaching them to be humane."—*Boston Sunday Globe.*

"A decided improvement on the general run of natural histories for young people." —*Morning Post.*

"Young people with a taste for natural history will be delighted with its pages, and we can strongly recommend it for either a prize or an addition to the library."— *School Newspaper.*

"An excellent little book."—*Daily News.*

"A very readable book; interesting and amusing, and, as it is meant to be, instructive. The illustrations will please young readers."—*Athenæum.*

"A capital natural history book."—*Graphic.*

"Crammed with good stories."—*Sheffield Independent.*

BOYS IN THE MOUNTAINS AND ON THE PLAINS; or, The Western Adventures of Tom Smart, Bob Edge, and Peter Small. By W. H. RIDEING, Member of the Geographical Surveys under Lieutenant Wheeler. With 101 Illustrations. Square 8vo. Cloth, gilt side and back, $2.50.

"A thoroughly fine book for boys in every way—full of interest and exciting adventures, crammed with information, physical and geographical, of the great West, its new life, its native tribes, its abundant animal life, its vast resources. We know of no other book in which all these matters are woven together in so attractive a form."— *The Churchman, New York.*

"The three boys of the story, which is not a compilation, but a record of Mr. Rideing's own observations, go through Colorado, Utah, and California, viewing the mountains, buttes, cañons, and lakes, and mixing with the inhabitants of all degrees of wildness. The many engravings of natural scenery are copied from photographs." —*Cincinnati Commercial.*

[SEE NEXT PAGE.]

BOOKS FOR YOUNG READERS.

THE YOUNG PEOPLE OF SHAKESPEARE'S DRAMAS. For Youthful Readers. By AMELIA E. BARR. With Illustrations. 12mo. Cloth, $1.50.

This work consists of scenes selected from Shakespeare's plays, in which youthful characters appear, accompanied with explanatory comments, and following each selection is an historic sketch, enabling the reader to compare the historical facts with the Shakespearean version. It is well calculated to please young readers.

"It was certainly a very happy thought which led Mrs. Barr to plan and execute this excellent book. Shakespeare has been drawn upon in many ways and by many writers, but the marshaling in line of his young people for the sake of young readers is, if we are not mistaken, breaking ground in a new direction. These children of Shakespeare have each of them something of the genius of their father, and to know them is to know something of the greatest mind in English literature; such an acquaintance can not be made too early or become too intimate, and, if Mrs. Barr's attractive book shall serve to interest young readers in the great dramatist, she will have performed a task of the most useful and beneficent kind."—*New York Christian Union.*

"A beautifully illustrated volume, and well fitted to lead to an interest in the masterpieces of English literature."—*New York Observer.*

"A novel and useful compilation."—*New York Independent.*

A WORLD OF WONDERS; or, Marvels in Animate and Inanimate Nature. A Book for Young Readers. With Three Hundred and Twenty-two Illustrations on Wood. Large 12mo. Cloth, illuminated, $2.00.

CONTENTS: Wonders in Marine Life; Curiosities of Vegetable Life; Curiosities of the Insect and Reptile World; Marvels of Bird and Beast Life; Phenomenal Forces of Nature.

"'A World of Wonders' reproduces for youthful learners in natural history a wide array of marvels from every department of the science. The curious growths of the seas and rivers, remarkable plants, and wonderful trees; singular insects and their still more singular instincts; birds of strange form and plumage; serpents and reptiles; striking phenomena of the air and water, ice, and fire, are all set forth with brief and simple descriptions and an abundance of excellent pictures which will take the attention of the most indifferent."—*Home Journal.*

A GEOGRAPHICAL READER, a Collection of Geographical Descriptions and Explanations, from the best Writers in English Literature. Classified and arranged to meet the wants of Geographical Students. By JAMES JOHONNOT. 12mo. Cloth, $1.25.

This volume has been compiled to furnish thought-reading to pupils while engaged upon the study of geography. It consists of selections from the works of well-known travelers and writers upon geography.

"Mr. Johonnot has made a good book, which, if judiciously used, will stop the immense waste of time now spent in most schools in the study of geography to little purpose. The volume has a good number of appropriate illustrations, and is printed and bound in almost faultless style and taste."—*National Journal of Education.*

For sale by all booksellers; or sent by mail, post-paid, on receipt of price.

New York: D. APPLETON & CO., 1, 3, & 5 Bond Street.

A LIBRARY

OF THE MOST IMPORTANT

STANDARD WORKS ON EVOLUTION.

I.

Origin of Species by Means of Natural Selection, or the Preservation of Favored Races in the Struggle for Life. By CHARLES DARWIN, LL. D., F. R. S. New and revised edition, with Additions. 12mo. Cloth, $2.00.

"Personally and practically exercised in zoölogy, in minute anatomy, in geology, a student of geographical distribution, not in maps and in museums, but by long voyages and laborious collection; having largely advanced each of these branches of science, and having spent many years in gathering and sifting materials for his present work, the store of accurately-registered facts upon which the author of the 'Origin of Species' is able to draw at will is prodigious."—*Professor T. H. Huxley*.

II.

Variation of Animals and Plants under Domestication. By CHARLES DARWIN, LL. D., F. R. S. With Illustrations. Revised edition. 2 vols., 12mo. Cloth, $5.00.

"We shall learn something of the laws of inheritance, of the effects of crossing different breeds, and on that sterility which often supervenes when organic beings are removed from their natural conditions of life, and likewise when they are too closely interbred."—*From the Introduction*.

III.

Descent of Man, and Selection in Relation to Sex. By CHARLES DARWIN, LL. D., F. R. S. With many Illustrations. A new edition. 12mo. Cloth, $3.00.

"In these volumes Mr. Darwin has brought forward all the facts and arguments which science has to offer in favor of the doctrine that man has arisen by gradual development from the lowest point of animal life. Aside from the logical purpose which Mr. Darwin had in view, his work is an original and fascinating contribution to the most interesting portion of natural history."

IV.

On the Origin of Species; or, The Causes of the Phenomena of Organic Nature. By Professor T. H. HUXLEY, F. R. S. 12mo. Cloth, $1.00.

"Those who disencumber Darwinism of its difficulties, simplify its statements, relieve it of technicalities, and bring it so distinctly within the horizon of ordinary apprehension that persons of common sense may judge for themselves, perform an invaluable service. Such is the character of the present volume."—*From the Preface to the American edition*.

V.

Darwiniana. Essays and Reviews pertaining to Darwinism. By ASA GRAY, Fisher Professor of Natural History (Botany) in Harvard University. 12mo. Cloth, $2.00.

"Although Professor Gray is widely known in the world of science for his botanical researches, but few are aware that he is a pronounced and unflinching Darwinian. His contributions to the discussion are varied and valuable, and as collected in the present volume they will be seen to establish a claim upon the thinking world, which will be extensively felt and cordially acknowledged. These papers not only illustrate the history of the controversy, and the progress of the discussion, but they form perhaps the fullest and most trustworthy exposition of what is to be properly understood by 'Darwinism' that is to be found in our language. To all those timid souls who are worried about the progress of science, and the danger that it will subvert the foundations of their faith, we recommend the dispassionate perusal of this volume."—*The Popular Science Monthly.*

VI.

Heredity: A Psychological Study of its Phenomena, Laws, Causes, and Consequences. From the French of TH. RIBOT. 12mo. Cloth, $2.00.

"Heredity is that biological law by which all beings endowed with life tend to repeat themselves in their descendants: it is for the species what personal identity is for the individual. The physiological side of this subject has been diligently studied, but not so its psychological side. We propose to supply this deficiency in the present work."—*From the Introduction.*

VII.

Hereditary Genius: An Inquiry into its Laws and Consequences. By FRANCIS GALTON, F. R. S., etc. New and revised edition, with an American Preface. 12mo. Cloth, $2.00.

"The following pages embody the result of the first vigorous and methodical effort to treat the question in the true scientific spirit, and place it upon the proper inductive basis. Mr. Galton proves, by overwhelming evidence, that genius, talent, or whatever we term great mental capacity, follows the law of organic transmission—runs in families, and is an affair of blood and breed; and that a sphere of phenomena hitherto deemed capricious and defiant of rule is, nevertheless, within the operation of ascertainable law."—*From the American Preface.*

VIII.

The Evolution of Man. A Popular Exposition of the Principal Points of Human Ontogeny and Phylogeny. From the German of ERNST HAECKEL, Professor in the University of Jena. With numerous Illustrations. 2 vols., 12mo. Cloth, $5.00.

"In this excellent translation of Professor Haeckel's work, the English reader has access to the latest doctrines of the Continental school of evolution, in its application to the history of man."

IX.

The History of Creation; or, the Development of the Earth and its Inhabitants by the Action of Natural Causes. A Popular Exposition of the Doctrine of Evolution in General, and of that of Darwin, Goethe, and Lamarck in Particular. By ERNST HAECKEL, Professor in the University of Jena. The translation revised by Professor E. RAY LANKESTER. Illustrated with Lithographic Plates. 2 vols., 12mo. Cloth, $5.00.

"The book has been translated into several languages. I hope that it may also find sympathy in the fatherland of Darwin, the more so since it contains special morphological evidence in favor of many of the most important doctrines with which this greatest naturalist of our century has enriched science."—*From the Preface.*

A STANDARD EVOLUTION LIBRARY.

X.

Religion and Science. A Series of Sunday Lectures on the Rlatione of Natural and Revealed Religion, or the Truths revealed in Nature and Scripture. By JOSEPH LE CONTE, LL. D. 12mo. Cloth, $1.50.

XI.

Prehistoric Times, as illustrated by Ancient Remains and the Manners and Customs of Modern Savages. By Sir JOHN LUBBOCK, Bart. Illustrated. Entirely new revised edition. 8vo. Cloth, $5.00.

The book ranks among the noblest works of the interesting and important class to which it belongs. As a *résumé* of our present knowledge of prehistoric man, it leaves nothing to be desired. It is not only a good book of reference, but the best on the subject.

XII.

Winners in Life's Race; or, The Great Backboned Family. By ARABELLA B. BUCKLEY, author of "The Fairy-Land of Science" and "Life and her Children." With numerous Illustrations. 12mo. Cloth, gilt side and back, $1.50.

XIII.

Physics and Politics; or, Thoughts on the Application of the Principles of "Natural Selection" and "Inheritance" to Political Society. By WALTER BAGEHOT. 12mo. Cloth, $1.50.

XIV.

The Theory of Descent and Darwinism. By Professor OSCAR SCHMIDT. With 26 Woodcuts. 12mo. $1.50.

"The facts upon which the Darwinian theory is based are presented in an effective manner, conclusions are ably defended, and the question is treated in more compact and available style than in any other work on the same topic that has yet appeared. It is a valuable addition to the 'International Scientific Series.'"—*Boston Post.*

XV.

Outline of the Evolution Philosophy. By Dr. M. E. CAZELLES. Translated from the French, by the Rev. O. B. FROTHINGHAM; with an Appendix, by E. L. YOUMANS, M. D. 12mo. Cloth, $1.00.

"This unpretentious little work will, no doubt, be used by thousands to whom the publications of Mr. Herbert Spencer are inaccessible and those of Auguste Comte repellent, by reason of their prolixity and vagueness. In a short space Dr. Cazelles has managed to compress the whole outline and scope of Mr. Spencer's system, with his views of the doctrine of progress and law of evolution, and a clear view of the principles of positivism."—*Nature (London).*

XVI.

Principles of Geology; or, The Modern Changes of the Earth and its Inhabitants, considered as illustrative of Geology. By Sir CHARLES LYELL, Bart. Illustrated with Maps, Plates, and Woodcuts. A new and entirely revised edition. 2 vols. Royal 8vo. Cloth, $8.00.

The "Principles of Geology" may be looked upon with pride, not only as a representative of English science, but as without a rival of its kind anywhere. Growing in fullness and accuracy with the growth of experi-

A STANDARD EVOLUTION LIBRARY.

ence and observation in every region of the world, the work has incorporated with itself each established discovery, and has been modified by every hypothesis of value which has been brought to bear upon, or been evolved from, the most recent body of facts.

XVII.
Elements of Geology. A Text-Book for Colleges and for the General Reader. By JOSEPH LE CONTE, LL. D., Professor of Geology and Natural History in the University of California. Revised and enlarged edition. 12mo. With upward of 900 Illustrations. Cloth, $4.00.

XVIII.
Animal Life, as affected by the Natural Conditions of Existence. By KARL SEMPER, Professor of the University of Würzburg. With Maps and 100 Woodcuts. 12mo. Cloth, $2.00.

XIX.
Crayfish: An Introduction to the Study of Zoölogy. By Professor T. H. HUXLEY, F. R. S. With 82 Illustrations. 12mo. Cloth, $1.75.

XX.
Anthropology: An Introduction to the Study of Man and Civilization. By EDWARD B. TYLOR, F. R. S. With 78 Illustrations. 12mo. Cloth, $2.00.

"The students who read Mr. Tylor's book may congratulate themselves upon having obtained so easy, pleasant, and workman-like an introduction to a fascinating and delightful science."—*London Athenæum.*

XXI.
First Principles. By HERBERT SPENCER. Part I. The Unknowable. Part II. The Knowable. 1 vol., 12mo. $2.00.

XXII.
The Principles of Biology. By HERBERT SPENCER. 2 vols., 12mo. $4.00.

XXIII.
The Principles of Psychology. By HERBERT SPENCER. 2 vols., 12mo. $4.00.

XXIV.
The Principles of Sociology. By HERBERT SPENCER. 12mo. 2 vols. $4.00.

XXV.
The Data of Ethics. By HERBERT SPENCER. Being Part I, Vol. I, of "The Principles of Morality." 12mo. Cloth, $1.25.

XXVI.
Illustrations of Universal Progress. By HERBERT SPENCER. 12mo. Cloth, $2.00.

For sale by all booksellers; or, sent by mail, post-paid, on receipt of price.

New York: D. APPLETON & CO., 1, 3, & 5 Bond Street.

JELLY-FISH, STAR-FISH, AND SEA-URCHINS.

Being a Research on Primitive Nervous Systems.

By G. J. ROMANES, F. R. S.,
Author of "Mental Evolution in Animals," etc.

12mo. Cloth, $1.75.

"A profound research into the laws of primitive nervous systems conducted by one of the ablest English investigators. Mr. Romanes set up a tent on the beach and examined his beautiful pets for six summers in succession. Such patient and loving work has borne its fruits in a monograph which leaves nothing to be said about jelly-fish, star-fish, and sea-urchins. Every one who has studied the lowest forms of life on the sea-shore admires these objects. But few have any idea of the exquisite delicacy of their structure and their nice adaptation to their place in nature. Mr. Romanes brings out the subtile beauties of the rudimentary organisms, and shows the resemblances they bear to the higher types of creation. His explanations are made more clear by a large number of illustrations. While the book is well adapted for popular reading it is of special value to working physiologists."—*New York Journal of Commerce.*

"Six years have been consumed in these investigations and experiments, and the result is a condensed statement of probably all that is known at present concerning these curious and beautiful fishes."—*New York Evening Telegram.*

"A most admirable treatise on primitive nervous systems. The subject-matter is full of original investigations and experiments upon the animals mentioned as types of the lowest nervous developments."—*Boston Commercial Bulletin.*

"Dr. Romanes is above all clear in his statements; the general reader for whom he writes in this volume, as well as the zoölogist, can find pleasure and instruction in what he has to say. The highly nervous organisms of which he treats are so familiar outwardly to the people, that what an eminent scientist writes after careful and intelligent investigation will be received with much more practical interest than the same scrutiny with extinct monsters."—*Hartford Evening Post.*

"A curious and instructive study of a low order of animal life, to which even the young naturalists who go for the summer to the sea-shore will be attracted. The book is meant, however, for something more than entertainment. The complicated and beautiful structure of the little marine animals is explained and illustrated very cleverly, and in a way to increase the wonder and admiration of the reader."—*Philadelphia Evening Bulletin.*

"Mr. Romanes's latest book discourses of ingenious experiments and original discoveries, and opens a field that has hitherto remained an unknown land to the general reader."—*Boston Saturday Evening Gazette.*

"May be read with delight and profit by all studious persons."—*Boston Beacon.*

"Mr. George J. Romanes has already established a reputation as an exact and comprehensive naturalist, which his later work, 'Jelly-Fish, Star-Fish, and Sea-Urchins,' fully confirms. These marine animals are well known upon our coasts, and always interest the on-lookers. In this volume (one of the International Scientific Series) we have the whole story of their formation, existence, nervous system, etc., made most interesting by the simple and non-professional manner of treating the subject. Illustrations aid the text, and the professional student, the naturalist, all lovers of the rocks, woods, and shore, as well as the general reader, will find instruction as well as delight in the narrative."—*Boston Commonwealth.*

New York: D. APPLETON & CO., 1, 3, & 5 Bond Street.

Origin of Cultivated Plants.

By ALPHONSE DE CANDOLLE.

12mo. Cloth, $2.00.

"The copious and learned work of Alphonse de Candolle on the 'Origin of Cultivated Plants' appears in a translation as volume forty-eight of the 'International Scientific Series.' Any extended review of this book would be out of place here, for it is crammed with interesting and curious facts. At the beginning of the century the origin of most of our cultivated species was unknown. It now requires more than four hundred closely printed pages to sum up what is known or conjectured of this matter. Among his conclusions M. Candolle makes this interesting statement: 'In the history of cultivated plants I have noticed no trace of communication between the peoples of the Old and New Worlds before the discovery of America by Columbus.' Not only is this book readable, but it is of great value for reference."—*New York Herald.*

"If general and lasting usefulness is to be accepted as the test of meritorious achievement, it might do us no harm to remember that within the last two thousand years man has not won from nature by discovery and cultivation a single species of food staple comparable with wheat, rice maize, the potato, and the banana, some of which invaluable conquests can indeed be proved to have been made more than forty centuries ago. This is one of the suggestive facts contributed by botany to the history of the birthplace and evolution of civilization, and which are discussed in the light of the ripest and most accurate investigation by the well-known Genevan scientist, Alphonse de Candolle, in the 'Origin of Cultivated Plants.' Of the one hundred and twenty to one hundred and forty thousand species in the vegetable kingdom, man has detected properties sufficiently precious to make it worth while to cultivate the plants on a large scale, and for use in less than three hundred instances."—*New York Sun.*

"Not another man in the world could have written the book, and, considering both its intrinsic merits and the eminence of its author, it must long remain the foremost authority in this curious branch of science. Of the 247 plants here enumerated, 199 are from the Old World, 45 are American, and 3 unknown. Of these only 67 are of modern cultivation. Curiously, however, the United States, notwithstanding its extent and fertility, makes only the pitiful showing of gourds and the Jerusalem artichoke."—*Boston Literary World.*

"The volume, though not large, is evidently the result of great study and research. Botanists in all parts of the world, travelers, herbaria, works on botany, history, archæology, and philology, have been all called upon to contribute to the results. The author justly remarks in his preface that 'the knowledge of the origin of cultivated plants is interesting to agriculturists, to botanists, and even to historians and philosophers concerned with the dawnings of civilization.'"—*Albany Cultivator and Country Gentleman.*

"In a word, all that can be recorded of the plant kingdom is given in the four hundred and fifty and more pages of the volume, and a good index aids materially the readers. All flower-lovers can find substantial information about their favorites in the work."—*Boston Commonwealth.*

"It is an exhaustive treatise on the habitat and origin of all the useful plants known to medicine, cookery, and science. One is here informed that the turnips come from Siberia, that the strawberry grows wild in Europe from Lapland to the isles of Madeira and Greece, gooseberries come from the Caucasus, and wheat from the region of the Euphrates. Rice was known 2,800 years B. C. in a ceremony in which the Emperor of China plays a conspicuous part."—*Cincinnati Enquirer.*

New York: D. APPLETON & CO., 1, 3, & 5 Bond Street.

CPSIA information can be obtained
at www.ICGtesting.com
Printed in the USA
BVHW070233270919
559497BV00003B/123/P